Springer Proceedings in Physics

Volume 428

Indexed by Scopus

The series Springer Proceedings in Physics, founded in 1984, is devoted to timely reports of state-of-the-art developments in physics and related sciences. Typically based on material presented at conferences, workshops and similar scientific meetings, volumes published in this series will constitute a comprehensive up to date source of reference on a field or subfield of relevance in contemporary physics. Proposals must include the following:

- Name, place and date of the scientific meeting
- A link to the committees (local organization, international advisors etc.)
- Scientific description of the meeting
- List of invited/plenary speakers
- An estimate of the planned proceedings book parameters (number of pages/articles, requested number of bulk copies, submission deadline).

Please contact:

For Americas and Europe: Dr. Zachary Evenson; zachary.evenson@springer.com
For Asia, Australia and New Zealand: Dr. Loyola DSilva; loyola.dsilva@springer.com

Jun Luo · Vadim Milyukov · Ze-Bing Zhou ·
Hsien-Chi Yeh · Jianwei Mei
Editors

Proceedings of the 7th International Workshop on the TianQin Science Mission; 25–26 April 2024, Hong Kong SAR, China

Progress on the TianQin Science Mission and in Gravitational Wave Detection

Springer

Editors
Jun Luo
Sun Yat-sen University
Guangzhou, Guangdong, China

Vadim Milyukov
Lomonosov Moscow State University
Moscow, Russia

Ze-Bing Zhou
Huazhong University of Science and Technology
Wuhan, Hubei, China

Hsien-Chi Yeh
Sun Yat-sen University
Guangzhou, Guangdong, China

Jianwei Mei
Sun Yat-sen University
Guangzhou, Guangdong, China

ISSN 0930-8989 ISSN 1867-4941 (electronic)
Springer Proceedings in Physics
ISBN 978-981-96-7802-0 ISBN 978-981-96-7803-7 (eBook)
https://doi.org/10.1007/978-981-96-7803-7

This Springer imprint is published by the registered company Springer Nature Singapore Pte Ltd.
The registered company address is: 152 Beach Road, #21-01/04 Gateway East, Singapore 189721, Singapore

Organizing Bodies of the Seventh International Workshop on the TianQin Science Mission

Scientific Advisory Committee

Masaki Ando
Rong-Gen Cai
Gerhard Heinzel
Martin A. Hendry
Sang Pyo Kim
Jun Luo
Vadim Milyukolv
Soumya D. Mohanty
Stefano Vitale
Bin Wang
Hsien-Chi Yeh
Ze-Bing Zhou

Local Organizers

Ming-chung Chu
Otto Hannuksela
Kenny Ng
Alejandro Torres Orjuela

Preface

The International Workshops on the TianQin Science Mission is a series that started in 2014. The workshops focus on the TianQin project and all related fields such as gravitational wave science, data analysis, multi-messenger astronomy, other current and future gravitational wave experiments, and recent observational results.

On the occasion of the 10-year anniversary of the proposal of the TianQin project, the 7th International Workshop on the TianQin Science Mission was hosted by Institute of Theoretical Physics, The Chinese University of Hong Kong from April 25 to 26, 2024. This event attracted over 130 experts and students from various countries, including China, Russia, Germany, the USA, Japan, and South Korea. The workshop included one keynote address, 30 plenary presentations, and 50 poster exhibits, providing a comprehensive platform for the exchange of cutting-edge research and innovative ideas in the field.

Guangzhou, China	Jun Luo
Moscow, Russia	Vadim Milyukov
Wuhan, China	Ze-Bing Zhou
Guangzhou, China	Hsien-Chi Yeh
Guangzhou, China	Jianwei Mei

Contents

Contributors

Yanzheng Bai National Gravitation Laboratory, MOE Key Laboratory of Fundamental Physical Quantities Measurement, and School of Physics, Huazhong University of Science and Technology, Wuhan, People's Republic of China

Yanzheng Bai National Gravitation Laboratory, MOE Key Laboratory of Fundamental Physical Quantities Measurement and School of Physics, Huazhong University of Science and Technology, Wuhan, China

Cosimo Bambi Center for Astronomy and Astrophysics, Center for Field Theory and Particle Physics and Department of Physics, Fudan University, Shanghai, People's Republic of China;
School of Humanities and Natural Sciences, New Uzbekistan University, Tashkent, Uzbekistan

Lin Cai National Gravitation Laboratory, MOE Key Laboratory of Fundamental Physical Quantities Measurement, and School of Physics, Huazhong University of Science and Technology, Wuhan, China

Ilya Chekh Sternberg Astronomical Institute, Moscow, Russia

Bingxue Chen National Gravitation Laboratory, MOE Key Laboratory of Fundamental Physical Quantities Measurement and School of Physics, Huazhong University of Science and Technology, Wuhan, China

Hui Chen National Gravitation Laboratory, MOE Key Laboratory of Fundamental Physical Quantities Measurement, and School of Physics, Huazhong University of Science and Technology, Wuhan, People's Republic of China

Jie-Wen Chen National Gravitation Laboratory, MOE Key Laboratory of Fundamental Physical Quantities Measurement, Department of Astronomy and School of Physics, Huazhong University of Science and Technology, Wuhan, China;
National Time Service Center, Chinese Academy of Sciences, Xi'an, China;

Key Laboratory of Time and Frequency Primary Standards, Chinese Academy of Sciences, Xi'an, China

Liu-Yang Chen National Gravimetry Laboratory, MOE Key Laboratory of Fundamental Physical Quantities Measurement, and School of Physics, Huazhong University of Science and Technology, Wuhan, China;
MOE Key Laboratory of Fundamental Physical Quantities Measurement and Hubei Key Laboratory of Gravitation and Quantum Physics, PGMF and School of Physics, Huazhong University of Science and Technology, Wuhan, China

Ti Chen Nanjing University of Aeronautics and Astronautics, Nanjing, China

Liangyu Chu National Gravitation Laboratory, MOE Key Laboratory of Fundamental Physical Quantities Measurement and School of Physics, Huazhong University of Science and Technology, Wuhan, China

Huizong Duan MOE Key Laboratory of TianQin Mission, TianQin Research Center for Gravitational Physics and School of Physics and Astronomy, Frontiers Science Center for TianQin, CNSA Research Center for Gravitational Waves, Sun Yat-sen University (Zhuhai Campus), Zhuhai, People's Republic of China

Sijun Fang MOE Key Laboratory of TianQin Mission, TianQin Research Center for Gravitational Physics & School of Physics and Astronomy, Frontiers Science Center for TianQin, Gravitational Wave Research Center of CNSA, Sun Yat-sen University (Zhuhai Campus), Zhuhai, China

Wei Feng School of Geospatial Engineering and Science, Sun Yat-sen University, Guangdong Zhuhai, China

Wen-Fan Feng National Gravitation Laboratory, MOE Key Laboratory of Fundamental Physical Quantities Measurement, Department of Astronomy and School of Physics, Huazhong University of Science and Technology, Wuhan, China;
Kavli Institute for Astronomy and Astrophysics, Peking University, Beijing, China

Yu-Jie Feng National Gravimetry Laboratory, MOE Key Laboratory of Fundamental Physical Quantities Measurement, and School of Physics, Huazhong University of Science and Technology, Wuhan, China;
MOE Key Laboratory of Fundamental Physical Quantities Measurement and Hubei Key Laboratory of Gravitation and Quantum Physics, PGMF and School of Physics, Huazhong University of Science and Technology, Wuhan, China

Yao Fu National Gravitation Laboratory, MOE Key Laboratory of Fundamental Physical Quantities Measurement, Department of Astronomy and School of Physics, Huazhong University of Science and Technology, Wuhan, China

Xida Han MOE Key Laboratory of TianQin Mission, TianQin Research Center for Gravitational Physics, School of Physics and Astronomy, Frontiers Science Center for TianQin Gravitational Wave Research Center, CNSA, Sun Yat-sen University (Zhuhai Campus), Zhuhai, China

Wei Hong National Gravitation Laboratory, MOE Key Laboratory of Fundamental Physical Quantities Measurement and School of Physics, Huazhong University of Science and Technology, Wuhan, China

Yi-Ming Hu MOE key Laboratory of TianQin Mission, TianQin Research Center for Gravitational Physics and Astronomy, Frontiers Science Center for TianQin, Gravitational Wave Research Center for CNSA, Sun Yat-sen University (Zhuhai Campus), Zhuhai, China;
School of Physics and Astronomy, Frontiers Science Center for TianQin, CNSA Research Center for Gravitational Waves, Sun Yat-sen University (Zhuhai Campus), Zhuhai, China

Xiangqing Huang MOE Key Laboratory of TianQin Mission, TianQin Research Center for Gravitational Physics and School of Physics and Astronomy, Frontiers Science Center for TianQin, CNSA Research Center for Gravitational Waves, Sun Yat-sen University (Zhuhai Campus), Zhuhai, People's Republic of China

Yuan-Ze Jiang National Gravimetry Laboratory, MOE Key Laboratory of Fundamental Physical Quantities Measurement, and School of Physics, Huazhong University of Science and Technology, Wuhan, China;
MOE Key Laboratory of Fundamental Physical Quantities Measurement and Hubei Key Laboratory of Gravitation and Quantum Physics, PGMF and School of Physics, Huazhong University of Science and Technology, Wuhan, China

Hong-Ming Jin National Gravitation Laboratory, MOE Key Laboratory of Fundamental Physical Quantities Measurement, Department of Astronomy and School of Physics, Huazhong University of Science and Technology, Wuhan, China

Bohong Li MOE Key Laboratory of TianQin Mission, TianQin Research Center for Gravitational Physics and School of Physics and Astronomy, Frontiers Science Center for TianQin, Gravitational Wave Research Center of CNSA, Sun Yat-sen University (Zhuhai Campus), Zhuhai, China

Enkun Li MOE key Laboratory of TianQin Mission, TianQin Research Center for Gravitational Physics and Astronomy, Frontiers Science Center for TianQin, Gravitational Wave Research Center for CNSA, Sun Yat-sen University (Zhuhai Campus), Zhuhai, China

Honggang Li National Gravitation Laboratory, MOE Key Laboratory of Fundamental Physical Quantities Measurement and School of Physics, Huazhong University of Science and Technology, Wuhan, China

Ming Li MOE Key Laboratory of TianQin Mission, TianQin Research Center for Gravitational Physics, School of Physics and Astronomy, Frontiers Science Center for TianQin Gravitational Wave Research Center, CNSA, Sun Yat-sen University (Zhuhai Campus), Zhuhai, China

Qing Li MOE Key Laboratory of Fundamental Physical Quantities Measurement, Hubei Key Laboratory of Gravitation and Quantum Physics, PGMF and School of

Physics, Huazhong University of Science and Technology, Wuhan, People's Republic of China

Wencan Li MOE Key Laboratory of Fundamental Physical Quantities Measurement, Hubei Key Laboratory of Gravitation and Quantum Physics, PGMF and School of Physics, Huazhong University of Science and Technology, Wuhan, People's Republic of China

Xin Li MOE Key Laboratory of Fundamental Physical Quantities Measurement, Hubei Key Laboratory of Gravitation and Quantum Physics, PGMF and School of Physics, Huazhong University of Science and Technology, Wuhan, People's Republic of China

Xinyu Li MOE Key Laboratory of TianQin Mission, TianQin Research Center for Gravitational Physics and School of Physics and Astronomy, Frontiers Science Center for TianQin, Gravitational Wave Research Center of CNSA, Sun Yat-sen University (Zhuhai Campus), Zhuhai, China

Xue-Yuan Li TianQin Research Center for Gravitational Physics and School of Physics and Astronomy, Sun Yat-sen University (Zhuhai Campus), Zhuhai, People's Republic of China

Yu-Rong Liang National Gravimetry Laboratory, MOE Key Laboratory of Fundamental Physical Quantities Measurement, and School of Physics, Huazhong University of Science and Technology, Wuhan, China;
MOE Key Laboratory of Fundamental Physical Quantities Measurement and Hubei Key Laboratory of Gravitation and Quantum Physics, PGMF and School of Physics, Huazhong University of Science and Technology, Wuhan, China

Fangjie Liao MOE Key Laboratory of TianQin Mission, TianQin Research Center for Gravitational Physics and School of Physics and Astronomy, Frontiers Science Center for TianQin, CNSA Research Center for Gravitational Waves, Sun Yat-sen University (Zhuhai Campus), Zhuhai, People's Republic of China

Hui Liu Harbin Institute of Technology, Harbin, China

Qi Liu TianQin Research Center for Gravitational Physics and School of Physics and Astronomy, Sun Yat-sen University (Zhuhai Campus), Zhuhai, People's Republic of China

Tan Liu National Gravitation Laboratory, MOE Key Laboratory of Fundamental Physical Quantities Measurement, Department of Astronomy and School of Physics, Huazhong University of Science and Technology, Wuhan, China;
School of Fundamental Physics and Mathematical Sciences, Hangzhou Institute for Advanced Study, UCAS, Hangzhou, China;
University of Chinese Academy of Sciences, Beijing, China

Yongqi Liu MOE Key Laboratory of Fundamental Physical Quantities Measurement, Hubei Key Laboratory of Gravitation and Quantum Physics, PGMF and School of Physics, Huazhong University of Science and Technology, Wuhan, People's Republic of China

Kian Hong Low School of Aerospace Engineering, Tsinghua University, Beijing, China

Jian Luo MOE Key Laboratory of TianQin Mission, TianQin Research Center for Gravitational Physics and School of Physics and Astronomy, Frontiers Science Center for TianQin, Gravitational Wave Research Center of CNSA, Sun Yat-sen University (Zhuhai Campus), Zhuhai, China

Jun Luo MOE Key Laboratory of TianQin Mission, TianQin Research Center for Gravitational Physics, School of Physics and Astronomy, Frontiers Science Center for TianQin, Gravitational Wave Research Center of CNSA, Sun Yat-sen University (Zhuhai Campus), Zhuhai, China

Xiangyu Lyu MOE key Laboratory of TianQin Mission, TianQin Research Center for Gravitational Physics and Astronomy, Frontiers Science Center for TianQin, Gravitational Wave Research Center for CNSA, Sun Yat-sen University (Zhuhai Campus), Zhuhai, China

Bo Ma School of Physics and Astronomy, Sun Yat-sen University, Zhuhai, China; CSST Science Center in the Great Bay Area, Zhuhai, China

Vadim Milyukov Lomonosow Moscow State University, Sternberg Astronomical Institute, Moscow, Russia;
Frontiers Science Center for TianQin, Sun Yat-sen University (Zhuhai Campus), Zhuhai, China

Soumya D. Mohanty Department of Physics and Astronomy, University of Texas Rio Grande Valley, Brownsville, TX, USA;
Department of Physics, IIT Hyderabad, Kandai, Telangana, India

Xiang Niu Harbin Institute of Technology, Harbin, China

Dan Peng TianQin Research Center for Gravitational Physics and School of Physics and Astronomy, Sun Yat-sen University (Zhuhai Campus), Zhuhai, People's Republic of China

Konstantin Postnov Sternberg Astronomical Institute, Moscow, Russia

Shao-Bo Qu National Gravitation Laboratory, MOE Key Laboratory of Fundamental Physical Quantities Measurement, and School of Physics, Huazhong University of Science and Technology, Wuhan, People's Republic of China

Rittick Roy Anton Pannekoek Institute for Astronomy, University of Amsterdam, Amsterdam, The Netherlands

Cheng-Gang Shao National Gravitation Laboratory, MOE Key Laboratory of Fundamental Physical Quantities Measurement, and School of Physics, Huazhong University of Science and Technology, Wuhan, People's Republic of China

Yong Shao School of Astronomy and Space Science, Nanjing University, Nanjing, China;
Key Laboratory of Modern Astronomy and Astrophysics, Nanjing University, Ministry of Education, Nanjing, China

Swarnim Shashank Center for Astronomy and Astrophysics, Center for Field Theory and Particle Physics and Department of Physics, Fudan University, Shanghai, People's Republic of China

Zhengyuan Shen National Gravitation Laboratory, MOE Key Laboratory of Fundamental Physical Quantities Measurement, and School of Physics, Huazhong University of Science and Technology, Wuhan, China

Yu-Jie Tan MOE Key Laboratory of Fundamental Physical Quantities Measurement and Hubei Key Laboratory of Gravitation and Quantum Physics, PGMF and School of Physics, Huazhong University of Science and Technology, Wuhan, China

Alejandro Torres-Orjuela Department of Physics, The University of Hong Kong, Pokfulam Road, Hong Kong;
Beijing Institute of Mathematical Sciences and Applications, Beijing, China

Changqing Wang State Key Laboratory of Geodesy and Earth's Dynamics, Innovation Academy for Precision Measurement Science and Technology, Chinese Academy of Sciences, Wuhan, China

Pan-Pan Wang National Gravitation Laboratory, MOE Key Laboratory of Fundamental Physical Quantities Measurement, and School of Physics, Huazhong University of Science and Technology, Wuhan, People's Republic of China

Yan Wang National Gravitation Laboratory, MOE Key Laboratory of Fundamental Physical Quantities Measurement, Department of Astronomy and School of Physics, Huazhong University of Science and Technology, Wuhan, China

Yankai Wang Nanjing University of Aeronautics and Astronautics, Nanjing, China

Zhaokui Wang School of Aerospace Engineering, Tsinghua University, Beijing, China

Zihan Wang MOE Key Laboratory of TianQin Mission, TianQin Research Center for Gravitational Physics and School of Physics and Astronomy, Frontiers Science Center for TianQin, Gravitational Wave Research Center of CNSA, Sun Yat-sen University (Zhuhai Campus), Zhuhai, China

Qian Wen School of Aerospace Engineering, Tsinghua University, Beijing, China

Jia-Qi Wu TianQin Research Center for Gravitational Physics and School of Physics and Astronomy, Sun Yat-sen University (Zhuhai Campus), Zhuhai, People's Republic of China

Ming-Yang Xu MOE Key Laboratory of Fundamental Physical Quantities Measurement and Hubei Key Laboratory of Gravitation and Quantum Physics, PGMF and School of Physics, Huazhong University of Science and Technology, Wuhan, China

Zhi-Lin Xu National Gravimetry Laboratory, MOE Key Laboratory of Fundamental Physical Quantities Measurement, and School of Physics, Huazhong University of Science and Technology, Wuhan, China

Xing-Yu Yan National Gravitation Laboratory, MOE Key Laboratory of Fundamental Physical Quantities Measurement, and School of Physics, Huazhong University of Science and Technology, Wuhan, People's Republic of China

Yihao Yan Max-Planck-Institut Für Gravitationsphysik, Albert-Einstein-Institut) and Institut Für Gravitationsphysik, Leibniz Universität Hannover, Hannover, Germany

Yong Yan MOE Key Laboratory of TianQin Mission, TianQin Research Center for Gravitational Physics and School of Physics and Astronomy, Frontiers Science Center for TianQin, Gravitational Wave Research Center of CNSA, Sun Yat-sen University (Zhuhai Campus), Zhuhai, China

Lingling Yang MOE Key Laboratory of Fundamental Physical Quantities Measurement, Hubei Key Laboratory of Gravitation and Quantum Physics, PGMF and School of Physics, Huazhong University of Science and Technology, Wuhan, People's Republic of China

Changqing Ye School of Physics and Electronic Engineering, Sichuan University of Science & Engineering, Zigong, People's Republic of China

Hsien-Chi Yeh MOE Key Laboratory of TianQin Mission, TianQin Research Center for Gravitational Physics and School of Physics and Astronomy, Frontiers Science Center for TianQin, CNSA Research Center for Gravitational Waves, Sun Yat-sen University (Zhuhai Campus), Zhuhai, People's Republic of China

Heng Yin MOE Key Laboratory of TianQin Mission, TianQin Research Center for Gravitational Physics and School of Physics and Astronomy, Frontiers Science Center for TianQin, Gravitational Wave Research Center of CNSA, Sun Yat-sen University (Zhuhai Campus), Zhuhai, China

Daren yu Harbin Institute of Technology, Harbin, China

Chunyu Yuan MOE Key Laboratory of TianQin Mission, TianQin Research Center for Gravitational Physics and School of Physics and Astronomy, Frontiers Science Center for TianQin, Gravitational Wave Research Center of CNSA, Sun Yat-sen University (Zhuhai Campus), Zhuhai, China

Jingyi Zhang MOE Key Laboratory of TianQin Mission, TianQin Research Center for Gravitational Physics and School of Physics and Astronomy, Frontiers

Science Center for TianQin, CNSA Research Center for Gravitational Waves, Sun Yat-sen University (Zhuhai Campus), Zhuhai, People's Republic of China

Hongchao Zhao MOE Key Laboratory of TianQin Mission, TianQin Research Center for Gravitational Physics and School of Physics and Astronomy, Frontiers Science Center for TianQin, Gravitational Wave Research Center of CNSA, Sun Yat-sen University (Zhuhai Campus), Zhuhai, China

Min Zhong School of Geospatial Engineering and Science, Sun Yat-sen University, Guangdong Zhuhai, China

Hongbo Zhou Harbin Institute of Technology, Harbin, China

Lixiang Zhou MOE Key Laboratory of TianQin Mission, TianQin Research Center for Gravitational Physics, School of Physics and Astronomy, Frontiers Science Center for TianQin Gravitational Wave Research Center, CNSA, Sun Yat-sen University (Zhuhai Campus), Zhuhai, China

Ze-Bing Zhou National Gravitation Laboratory, MOE Key Laboratory of Fundamental Physical Quantities Measurement, and School of Physics, Huazhong University of Science and Technology, Wuhan, People's Republic of China

Fan Zhu MOE Key Laboratory of TianQin Mission, TianQin Research Center for Gravitational Physics and School of Physics and Astronomy, Frontiers Science Center for TianQin, CNSA Research Center for Gravitational Waves, Sun Yat-sen University (Zhuhai Campus), Zhuhai, People's Republic of China

Zitong Zhu State Key Laboratory of Geodesy and Earth's Dynamics, Innovation Academy for Precision Measurement Science and Technology, Chinese Academy of Sciences, Wuhan, China

Part I
Overview

Chapter 1
Ten Years of TianQin Project

Vadim Milyukov

Abstract This paper provides an overview of the development of the TianQin space mission as well as a concise history of gravity experiments conducted in China. Over the past decade, seven workshops on TianQin project have been held to discuss the project's findings and to delineate future plans. In order to facilitate international collaboration on the project and to enhance its visibility within the international scientific community, an International Advisory Committee of the TianQin Mission has been constituted.

1.1 Introduction (History of High Precision Gravity Measurements in China)

The history of fundamental gravitational measurements in China dates back to the mid-70 s of the last century. At Sun Yat-sen University, a group of researchers led by Prof. Jiayan Chen built a gravitational wave detector and began efforts to detect gravitational waves. This was a room-temperature bar antenna (also known as a Weber bar), a simple device for detecting the expected gravitational wave radiation, which was widely used around the world in the 1960s and 1970s. Unfortunately, attempts to detect gravitational waves using such detectors were unsuccessful, despite Weber's own claims of successful detection of gravitational waves [1].

Ten years later, in the mid-80 s of the last century, Luo Jun, a young graduate of Huazhong University of Science and Technology (at that time Huazhong Institute of Technology), became a lecturer at this university and began to be actively involved in scientific activities. Initially, his main scientific interest was the experimental determination of Newton's gravitational constant G.

V. Milyukov (✉)
Lomonosow Moscow State University, Sternberg Astronomical Institute, Moscow, Russia
e-mail: vmilyukov@yandex.ru

Frontiers Science Center for TianQin, Sun Yat-Sen University (Zhuhai Campus), Zhuhai, China

J. Luo et al. (eds.), *Proceedings of the 7th International Workshop on the TianQin Science Mission; 25–26 April 2024, Hong Kong SAR, China*, Springer Proceedings in Physics 428, https://doi.org/10.1007/978-981-96-7803-7_1

The traditional instrument for measuring the interaction of weak forces, particularly gravitational forces, is the torsion balance. The first experiment to determine the Newtonian gravitational constant with a nonlinear fitting method was performed at HUST in the late 1990s and was published in the authoritative journal Physical Review D in 1999 [2]. The obtained value of the gravitational constant with the 2005 correction [3] was included in the series of experimental values of G, upon which the CODATA values of the gravitational constant was derived in 2002 and 2006 (CODATA, is the Task Group on Fundamental Constants of the Committee on Data for Science and Technology).

The results of the next experiment to determine G were published in 2009 and 2010 [4, 5]. The new value of the Newtonian gravitational constant G was obtained by using the time-of-swing method. The method and technique of the previous experiment were improved, resulting in a significant reduction in the uncertainties of the gravitational constant to 26 ppm. The new value of the gravitational constant, designated as HUST-09 in the 2010 CODATA adjustment, was $G = (6.67349\ 0.00018) \times 10^{-11}\ m^3kg^{-1}\ s^{-2}$ and represented one of the most precise determinations at that time.

Despite the large number of experiments conducted to determine the gravitational constant in a variety of laboratories around the globe, the precise value of G remained elusive. In 2016 GODATA published a value of G with a relative error of 47×10^{-6}, which exceeds the precision of individual experiments. This considerable discrepancy indicates the potential presence of unidentified systematic errors across the various experiments. One potential solution to this problem is to measure G using a combination of different methods and experimental setups.

Two independent experiments were conducted at HUST. The gravitational constant was determined through the application of two distinct methodologies: the time-of-swing method and the angular acceleration feedback method. Two values of the gravitational constant, $G = 6.674184(78) \times 10^{-11}\ m^3kg^{-1}\ s^{-2}$ and $G = 6.674484(78) \times 10^{-11}\ m^3kg^{-1}\ s^{-2}$, were obtained with a relative error of 11.6×10^{-6}. These two values are in agreement with one another within the two-standard-deviation confidence level and also align with the value of G recommended by CODATA. The result obtained allows us to consider it highly probable that a value of the gravitational constant free from systematic errors has been measured. The accuracy achieved is the best in the world.

The results are published in Nature, 560, pp. 582–588 (2018) [6]. In announcing this publication, the journal wrote: “Gravity measured with record precision.”

A significant research area of the Center for Gravitational Experiments (CGE), established by Prof. Luo Jun at HUST, has been the verification of the inverse square law in the millimeter and submillimeter ranges. These experiments were initiated in laboratories around the world based on theoretical speculations predicting a violation of the Newtonian inverse square law (ISL) in the non-relativistic limit in a weak gravitational field. The deviations from the ISL are usually parameterized by an exponential Yukawa potential, which is incorporated into the standard gravitational potential and defined by three parameters: α (the strength of the new interaction, which is postulated to be a universal constant and independent of the composition), λ (the length scale), and r (the separation between two masses) [7].

In the series of experiments with successively increasing constraints on the parameters of the non-Newtonian interaction, the inverse square law was tested at millimeter and submillimeter ranges. The results of these experiments showed that the gravitational inverse square law holds with 95% confidence in the ranges from 0.7 to 5.0 mm and 40–350 μm [8–13]. These experiments, performed by CGE over a period of approximately fifteen years, have made a significant contribution to fundamental physics, confirming the validity of the General Relativity.

Another original experiment conducted with the rotating torsion balance was a test of the equivalence principle (one of the postulates of Einstein's general theory of relativity). This experiment was designed to search for a possible violation of gravitational parity using test bodies with different chirality. The results demonstrated that the difference of accelerations towards the Earth is not greater than $(-1.7 \pm 4.1) \times 10^{-15}\ \mathrm{ms}^{-2}$ (1-σ statistical uncertainty). This was the first experimental verification of the equivalence principle for chiral masses, which opened a new way to search for possible parity-violating gravity [14].

The scientific interests of CGE were not limited only to fundamental gravitational experiments. A wide range of experimental activities was also extended to other areas of physics. Notable achievements include the series of experiments on the limit of the rest mass of the photon [15, 16], the search for violations of Lorentz invariance [17], and the development of an atomic interferometer for gravitational measurement [18].

The HUST Center for Gravitational Experiments is currantly a world-class laboratory whose scientific achievements are known worldwide, extending far beyond China.

1.2 Chinese-Russian Relations

In the mid-1980s, fruitful scientific contacts were established between HUST and Lomonosov Moscow State University initially based on a mutual scientific interest in the experimental determination of the gravitational constant. Subsequently, with the appointment of Academician Luo Jun as President of SYSU, such contacts were also established between SYSU and MSU. During the period of our cooperation, a number of joint Chinese - Russian grants have been carried out under the cooperation of the National Science Foundation of China (NSFC) and the Russian Foundation for Basic Research (RFBR). At present, a joint Chinese - Russian grant "The TianQin Project: Study of the mission space environment and gravitational waves sources" is being implemented by the TianQin Research Center, Sun Yat-sen University and Sternberg Astronomical Institute of Lomonosov Moscow State University under the cooperation between the National Science Foundation of China and the Russian Science Foundation.

In 2017 in Moscow, in a solemn atmosphere, the President of SYSU Academician Jun Luo and the Rector of Moscow State University Academician Viktor Sadovnichy signed the Agreement on scientific, educational and cultural cooperation between SYSU and MSU. A separate Annex to this Agreement notes the cooperation of the two Universities in the field of gravitational wave experiments, including the TianQin project. In addition to this Agreement, the Cooperation Agreement was signed between SYSU and MSU on the TianQin-2 mission in 2020. With the entry into force of these two agreements, our scientific interaction gains additional opportunities.

During the Russian-Chinese meetings in 2013–2014, when discussing the prospects for the development of joint research, it was realized, that the capabilities of the Terrestrial Laboratory for a number of gravity experiments have already been exhausted. The laboratory, even in "ideal" conditions of underground location, with extremely low seismic noise, is clearly not sufficient to achieve high accuracy of measurement of fine gravitational effects. Access to a space laboratory is necessary. The development of space technologies provides new opportunities to measure the fine gravitational effects that underlie relativistic theories of gravity, including General Relativity. Placing instruments in space provides access to conditions of special dynamical "purity" that are unattainable in terrestrial laboratories, but which are crucial for the success of precision experiments. For this reason, space basing is essential for many experiments in the field of fundamental gravitation.

1.3 From the Terrestrial Laboratory to the Space Observatory (Concept of the TianQin Project)

From the point of view of experimental testing of gravitational theories, our solar system is a unique "laboratory" in which the conditions necessary for conducting the most important fundamental research are present (gravitational potentials near the Sun and in terrestrial conditions differ by a factor of 3,000). Nevertheless, evaluations show that precision gravity experiments are also possible in near-Earth orbits, and launching a satellite into Earth orbit is a much more affordable task than launching a satellite into solar orbit. Therefore, developing the concept of relativistic gravity experiments using a cluster of satellites in Earth orbit was recognized as a promising direction for our further research. An unprecedented level of accuracy in measuring gravitational effects will be provided by precision laser links capable of measuring inter-satellite range and range rate at the level of 1 pm and 1 pm/s, respectively.

1.3.1 Measuring the Post Newtonian Parameter Gamma

Initially, two space project concepts were considered at the Chinese -Russian workshop "Towards to Future Gravitation Space Mission" held in Wuhan in March 2014.

The first option was aimed at measuring the parameterized post-Newtonian parameter γ. The parameterized post-Newtonian formalism, or the PPN formalism, is represented by the parameters by which a General Theory of Gravity can differ from Newtonian gravity [19–22]. In general, PPN formalism can be applied to compare Einsteinian gravity and all metric theories of gravitation in which all bodies satisfy the Einstein equivalence principle (EEP) [23]. The PPN parameter γ (the Eddington parameter), whose value in General Relativity is unity, describes the space curvature g_{ij} produced by the unit rest mass. Additionally, geodetic precession (de Sitter effect) [24, 25] and Lense-Thirring precession (frame dragging) [26] in the Earth's gravitomagnetic field can be measured in such a project.

The concept for the measurement of the PPN parameter γ is based on the use of a cluster of 4 satellites placed in one plane on circular orbits of 80,000 km (corresponding orbital period ~ 2.6 days) in the Earth's gravitational field. The satellites are distributed along the orbit in the form of a cross and connected by a laser interferometer links. The laser beams form a trapezoid with diagonal elements $\approx$160,000 km, so that the light beam of one of the diagonals passes close to the Earth and experiences an additional delay due to the Earth's gravitational field (Shapiro effect) [27]. In Euclidean geometry, such a system is redundant: by measuring only five ranges, it is possible to calculate the sixth. However, the experiment will measure all 6 inter-satellite ranges in order to detect the difference between the geometry of space and Euclidean geometry and also avoid Earth-based range measurements, which has much lower accuracy.

The scientific goal of this space experiment is to measure the parameter γ with an unprecedented accuracy, of the order of 10^{-9}, i.e., 30,000 times more accurate than the best modern result obtained in 2002 by the microwave location of the CASSINI spacecraft ($\gamma - 1 = (2.1 \pm 2.3) \times 10^{-5}$) [28]. The scientific significance of improving the accuracy of the γ parameter is due to the fact that numerous models of scalar-tensor theories of gravity, consistent with modern cosmological observations, predict the lower limit of the variation of these parameters at the level of $\approx 10^{-6}$–10^{-7}. Thus, the improvement of the measurement of this parameter should provide crucial information to confirm the validity of either GR or scalar-tensor theories of gravitation, to check the modern theories of cosmological evolution.

1.3.2 Gravitational Wawe Detection

The second gravitational space mission concept considered in Wuhan in March 2014 was the concept of the Gravitational Wave Detector Project. The goal of the project is to place a constellation of satellites in geocentric orbit and record gravitational waves in the millihertz range (0.1–100 mHz). Gravitational wave radiation in this frequency range can be caused by many different astronomical sources, such as ultra-compact galactic binaries, mergers of massive black holes, the capture of stellar objects by a massive black hole, and others.

A distinctive aspect of the project currently under development is that the instruments, design, and data processing algorithms utilized in this space experiment will be optimized for the detection of gravitational waves (GWs) from a selected, most promising gravitational radiation source, referred to as the "calibration source." In accordance with this objective, the initial phase of the mission is focused on the development of a space-based detector of gravitational waves from a selected well-known source. The initial objective of the detector is to directly detect gravitational waves with the anticipated characteristics of the source in question. The detector, which is a cluster of satellites, can be fully controlled and utilized for detailed studies of the selected source. Subsequently, the detector will function as a gravitational-wave observatory, capable of detecting gravitational radiation from other sources, both known and unknown.

Additionally, the project was driven by the necessity to develop a space-based gravitational-wave detector within a limited timeframe, in response to the competitive landscape established by the LISA mission [29, 30]. The objective is to launch a constellation of satellites into orbit and to achieve reliable registration of GW radiation from the selected source within a timeframe of 18–20 years. An essential preliminary step toward the realization of this approach is a realistic assessment of the current state of technology, identification of technological gaps, and the initiation of efforts to eliminate them.

At the same workshop, Prof. Luo Jun suggested naming the project "TianQin", which means "Heavenly Citra". It is easy to see the poetic allegory in this name: the Universe makes our Heavenly Citra sound through graceful fluctuations of gravitational waves, creating a wonderful beauty of magical music…

The detailed design of the orbital experiment scheme showed that the space detector should be realized on three identical spacecraft placed on almost identical geocentric orbits with a large semi-major axis of ~ 10^5 km and forming an almost equilateral triangle. Compared to heliocentric orbits, the choice of more affordable geocentric orbits allows the use of more affordable technologies for the spacecraft, which significantly reduces the overall cost of the mission.

As a reference source, the most suitable candidate is the double star RX J0806.3 + 1527, due to the fact that it has the shortest period of known among double systems, its relative proximity to the Sun, and its moderate component masses. For this source and interferometer, with arm lengths $L \sim 10^5$ km, a signal-to-noise ratio on the order of 10 can be achieved within three months of observations.

A description of the TianQin mission concept, source selection, spacecraft orbit calculation, required sensitivity of the experiment, requirements for the two key instruments: the laser interferometer and the drag-free system, as well as error budget estimation of these key instrumental components, can be found in [31]. For a comprehensive overview on the current status of the TianQin project, please refer to the article by Luo Jun in these Proceedings.

1.4 TianQin Project Development and International Cooperation

The first Workshop on the TianQin space mission was held in Wuhan in December 2014. A total of 33 people attended the meeting. Seventeen people represented the hosts of the workshop, the HUST Center for Gravitational Experiments, led by Prof. Luo Jun. Five participants were from Russia, Moscow State University. Eleven people represented different organizations in China: seven from different institutes of CAS, two from DFH Satellite Co., LTD and two from Chinese universities. Thus, the first Workshop was already international in nature, with a fairly broad representation of Chinese science and space organizations.

This Workshop continued the discussion of two concepts for space gravity missions, but ultimately the TianQin gravitational wave detector project was recognized as the priority for further development and implementation.

The first TianQin mission Workshop focused on discussing and exploring the scientific objectives, mission concept, and key technologies required for the TianQin science mission. It was envisioned that a feasibility analysis and a detailed roadmap would be developed as the first steps of the project, and all scientists and researchers willing to participate in this important science mission would be assigned relevant research tasks, e.g. GW source studies, orbit design, satellite platform development, payload instrumentation.

In January 2015, the project leader, Academician Luo Jun, was appointed President of Sun Yat Sen University, and **the 2th Workshop** of the TianQin mission was held in Zhuhai in May 2015. About 40 participants attended the Workshop: 15 people from HUST and SYSU and 25 people from other scientific organizations in China and Russia (5 people). In addition to discussing the upcoming goals of project development, an important key decision was made at the meeting. It was decided to establish the TianQin Research Center at SYSY for the implementation of the GW detector project under the leadership of SYSY President Luo Jun. In the first stage, only a few people formed part of the Center. Among them were Professors Jianwei Mei and Hsien-Chi Ye. Formally staff members of HUST at that time, they played a key role in establishing the center and organizing its work in the initial stage.

The next workshop**, the 3th Workshop** on the TianQin mission, took place six months after the second workshop, in December 2015, in Guangzhou. In addition to Chinese participants, foreign participants were well represented at this meeting. A relatively large number of participants came from Russia: from Moscow State University, Lavochkin Space Corporation, and "Precision Systems and Instruments" Corporation. The latter two are part of the Russian State Space Corporation ROSCOSMOS. The workshop was also attended by representatives from other countries: Gerhard Heinzel (Germany), Viktor Toth (USA) and Enrico Canuto (Italy).

The conference lasted three days and was divided into five sessions: (1) System Simulation; (2) Disturbance Reduction System; (3) Science and More; (4) Laser and Satellite; (5) Space-Related Experiments.

From the titles of the sessions it can be seen that, in contrast to previous workshops where mainly conceptual and organizational issues were discussed, this workshop already presented concrete systematic development in all directions of the project implementation. A number of presentations can be cited as examples: (1) The Inertial Sensor for the TianQin Mission (Haibo Tu); (2) Thrusters for Drag-Free Control and Orbit Transfer (Jin-Yue Geng); (3) Preliminary Definition of the TianQin Satellite (Yong-He Zhang); (4) The Prospect of a CCR on the Change'4 Relay Satellite (Hsien-Chi Ye). Reports were also given by representatives of MSU, Lavochkin Sapce Corporation, "Precision Systems and Instruments" Corporation, and other foreign participants.

The 4th Workshop on the TianQin project was held in Whuhan in November 2016. The main topics discussed at this meeting were related to the current progress in the development of the key technologies of the TianQin project: the laser interferometer and the inertial sensor. The fourth workshop was also dedicated to the sixtieth anniversary of the project leader, Academician Jun Luo.

In December 2018, **the 5th Workshop** on the TianQin science mission was held in Zhuhai. The purpose of this workshop was to discuss the tasks of the TianQin project in the coming years, based on the work done in the past few years. In order to facilitate more and more domestic and foreign experts to participate in the detailed work of the TianQin project, the TianQin Collaboration was established during this workshop. The core objective of the TianQin Collaboration is to actively promote the work of the TianQin Project by introducing an efficient form of organization and management.

In this regard, about fifteen bilateral Agreements of Cooperation between TianQin Research Center and various scientific and industrial organizations from across China were signed in a solemn ceremony at the beginning of the workshop.

At the same workshop, the general structure of the TianQin project was approved. According to this structure, the project is implemented at Sun Yat-Sen University under the leadership of the Project Leader, Academician Luo Jun. The Executive Board, consisting of the Executive Leader, Management and Working Group Leaders, manage the practical work of the project. The working groups carry out the work in the main areas of the project. The main activities of the working groups are scientific (Science, Data Analysis, Education, etc.) and technological (Laser, Inertial Sensor, Satellite, etc.).

In order to develop international cooperation on the project, the International Advisory Committee of the TianQin Mission (TQIAC) has been established. The main function of TQIAC is to provide an external view on the development of the project and to promote the TianQin project in the international scientific community. The initial membership of TQIAC consisted of 7 foreign scientists, including well-known experts in the field of gravitational wave experiment David Blair (Australia), Stefano Vitale (Italy), Gerhard Heinzel (Germany). The TQIAC was chaired by Vadim Milyukov (Russia).

In December 2019, **the 6th Workshop** of the TianQin project was held in Whuhan. This meeting was attended by a record number of participants, a few hundred, most of whom were students. Thus, this meeting fulfilled not only a scientific function but also had an educational value.

After more than four years, **the 7th Workshop** on the TianQin mission was held in April 2024. For the first time, it was held in Hong Kong rather than in mainland China. One of the motivations for holding it in Hong Kong was to make it more accessible to overseas participants. Over the two days of the meeting, more than 40 papers were presented by both Chinese and foreign participants. The keynote talks focused on the TianQin project and all related areas such as gravitational wave science, data analysis, multi-messenger astronomy, other current and future gravitational wave experiments, and recent observational results. These Proceedings contain most of the papers presented at the 7th TianQin Workshop.

The TQIAC meeting was also held at the 7th Workshop. The membership of the commission was renewed. TQIAC sees its future activities in the following directions:

To help steer the scientific development of TianQin through advisory function;
Improving the international visibility of TianQin through advocacy;
Promoting international cooperation on TianQin, including both science and technology, through collaborative research, joint projects, training and education.

Two additional events related to international collaboration are worthy of mention.

In September 2018, **the Academic Seminar No. S45**, held within the framework of the Xiangshan Science Conferences (XSSC), addressed the topic of "The TianQin Project and International Collaboration." The XSSC seminars are organized by the Ministry of Science and Technology of China with the support of the Chinese Academy of Sciences and other Chinese organizations. The XSSC promotes scientific research, academic exchange, and the prioritization of scientific frontiers and their future development. The seminar was helmed by Executive Chairs Academicians of the Chinese Academy of Sciences, Rong-Gen Cai and Jun Luo, as well as Professor Vadim Milyukov from Moscow State University. The seminar was attended by eminent foreign scientists in the domain of gravity and gravitational wave experiments. The seminar addressed a wide range of issues, both scientific and technological, related to the future development of the TianQin project.

In 2023, the Ministry of Education of China (MOEC) has approved the establishment of the **"Frontiers Science Center for TianQin"** (FSCTQ). The center is based on the existing TianQin center and serves as a means of providing more stable support for the TianQin project. The center is governed by an Academic Committee, whose responsibilities include establishing the center's direction of development, facilitating academic exchanges, and assisting the center in recruiting world-class researchers. The committee is comprised of experts from both China and abroad.

1.5 Conclusion

Today we summarize the results of creative work that has been conducted on the development of the TianQin project. Formally, half of the way has been passed: let us remember that the project was originally planned to be completed within 20 years. Over the past 10 years, significant progress has been made. Precision technology of scientific instruments is being created; scientific objectives of the mission are being developed in detail; algorithms and software for processing and analyzing data of the future GW mission are being implemented; technologies for launching and orbiting TianQin satellites are being developed; and finally, two stages of the mission roadmap have been implemented. Detailed information on the results obtained can be found in the articles in this Proceedings. For comprehensive information on the current status of the TianQin project implementation, please refer to Luo Jun's article in this same volume, so I will not dwell on specific scientific results.

It is worth noting the outstanding contributions of Project leader, Academician Luo Jun, in the conceptualization, establishment, and advancement of the TianQin Project. His role exemplifies not only his expertise in scientific and organizational leadership but also his proficiency in pedagogy. Over the span of four decades, he has mentored numerous highly skilled professionals, many of whom are currently engaged in the TianQin Project and thanks to whom it is successfully developing.

The second half of the work to realize the TianQin mission is forthcoming and may prove to be the most challenging. However, the work that has already been completed, along with the rate at which it has been implemented, provides assurance that the TianQin mission will be accomplished and that the scientific objectives of the mission will be achieved.

Acknowledgements The author acknowledge the support from the Russian Science Foundation grant No 23-42-00055.

References

1. J. Weber, Evidence for discovery of gravitational radiation. Phys. Rev. Lett. **22**(24), 1320 (1969)
2. J. Luo, Z.-K. Hu, X.-H. Fu, S.-H. Fan, M.-X. Tang, Determination of the Newtonian gravitational constant G with a nonlinear fitting method. Phys. Rev. D. **59**, 042001 (1999)
3. Z.-K. Hu, J.-Q. Guo, J. Luo, Correction of source mass effects in the HUST-99 measurement of G. Phys. Rev. D **71**, 127505 (2005)
4. J. Luo, Qi. Liu, L.-C. Tu, C.-G. Shao, L.-X. Liu, S.-Q. Yang, Q. Li, Y.-T. Zhang, Determination of the Newtonian gravitational constant G with time-of-swing method. Phys. Rev. Lett. **102**, 240801 (2009)
5. L.-C. Tu et al., New determination of the gravitational Constant G with time-of-swing method. Phys. Rev. D **82**, 022001 (2010)
6. Q. Li, C. Xue, J.-P. Liu, J.-F. Wu, S.-Q. Yang, C.-G. Shao, L.-D. Quan, W.-H. Tan, L.-C. Tu, Qi. Liu, Xu. Hao, L.-X. Liu, Q.-L. Wang, Z.-K. Hu, Z.-B. Zhou, P.-S. Luo, S.-C. Wu, V. Milyukov, J. Luo, Measurements of the gravitational constant using two independent methods. Nature **560**, 582–588 (2018)

7. E.G. Adelberger, B.R. Heckel, A.E. Nelson, Tests of the gravitational inverse-square law. Annu. Rev. Nucl. Part. Sci. **53**, 77 (2003)
8. S.-Q. Yang, B.-F. Zhan, Q.-L. Wang, C.-G. Shao, L.-C. Tu, W.-H. Tan, J. Luo, Test of the gravitational inverse square law at millimeter ranges. Phys. Rev. Lett. **108**, 081101 (2012)
9. L.-C. Tu, S.-G. Guan, J. Luo, C.-G. Shao, L.-X. Liu, Null test of Newtonian inverse-square law at submillimeter range with a dual-modulation torsion pendulum. Phys. Rev. Lett. **98**, 201101 (2007)
10. S.-G. Guan, L.-C. Tu, J. Luo, C.-G. Shao, L.-X. Liu, Null test of Newtonian inverse-square-law at short range: current status and developments. Prog. Theor. Phys. Suppl. (172), 192–201 (2008)
11. W.-H. Tan, S.-Q. Yang, C.-G. Shao, J. Li, D. An-Bin, B.-F. Zhan, Q.-L. Wang, P.-S. Luo, L.-C. Tu, J. Luo, New test of the gravitational inverse-square law at the submillimeter range with dual modulation and compensation. Phys. Rev. Lett. **116**, 131101 (2016)
12. J. Wang, S. Guan, K. Chen, W. Wu, Z. Tian, P. Luo, A. Jin, S. Yang, C. Shao, J. Luo, Test of non-Newtonian gravitational forces at micrometer range with two dimensional force mapping. Phys. Rev. D **94**, 122005 (2016)
13. W.-H. Tan, A.-B. Du, W.-C. Dong, S.-Q. Yang, C.-G. Shao, S.-G.-L. Wang, B.-F. Zhan, P.-S. Luo, L.-C. Tu, J. Luo, Improvement for testing the gravitational inverse-square law at the submillimeter range. Phys. Rev. Lett. **124**, 051301 (2020)
14. L. Zhu, Q. Liu, H.-H. Zhao, Q.-L. Gong, S.-Q. Yang, P. Luo, C.-G. Shao, Q.-L. Wang, T. Liang-Cheng, J. Luo, Test of the equivalence principle with chiral masses using a rotating torsion pendulum. Phys. Rev. Lett. **121**, 261101 (2018)
15. L.-C. Tu, J. Luo, Experimental tests of Coulomb's Law and the photon rest mass. Metrologia **41**, S136–S146 PII: S0026–1394 (04) 80308–6 (2004)
16. L.-C. Tu, J. Luo, G.T. Gillies, The mass of the photon. Rep. Prog. Phys. **68** (2005)
17. C.-G. Shao, Y.-J. Tan, W.-H. Tan, S.-Q. Yang, J. Luo, M.E. Tobar, Search for Lorentz invariance violation through tests of the gravitational inverse square law at short ranges. Phys. Rev. D **91**, 102007 (2015)
18. Z.H.O.U. Min-kang, H.U. Zhong-kun, D.U.A.N. Xiao-chun, S.U.N. Bu-liang, Z.H.A.O. Jin-bo, L.U.O. Jun, Experimental progress in gravity measurement with an atom interferometer. Front. Phys. China. **4**(2), 170–173 (2009)
19. A.S. Eddington, Mathematical theory of relativity. (Cambridge University Press, Cambridge, 1922) p. 270
20. C.W. Misner, K.S. Thorne, J.A. Wheeler, Gravitation. (Princeton University Press, 2017) p. 1336
21. C.M. Will, Theoretical frameworks for testing relativistic gravity. II. parametrized Post-Newtonian hydrodynamics, and the Nordtvedt effect. Astrophys. J. **163**, 611–628 (1971)
22. C.M. Will, K. Nordtvedt Jr, Conservation laws and preferred frames in relativistic gravity. I. Preferred-Frame Theories and an Extended PPN Formalism. Astrophys. J. **177**, 757 (1972)
23. K. Nordtvedt, Equivalence principle for massive bodies including rotational energy and radiation pressure. Phys. Rev. **180** (5), 1293 (1969)
24. W. De Sitter, On Einstein's Theory of gravitation and its astronomical consequences. Mon. Not. R. Astron. Soc. **77**, 155–184 (1916)
25. H. Pfister, On the history of the so-called Lense-Thirring effect. Gen Relativ Gravity. **39**, 1735–1748 (2007)
26. L. Iorio, Juno, the angular momentum of Jupiter and the Lense-Thirring effect. New Astronomy **15**(6), 554–560 (2010)
27. I.I. Shapiro, Fourth test of general relativity. Phys. Rev. Lett. **13**(26), 789–791 (1964)
28. B. Bertotti, L. Iess, P. Tortora, A test of general relativity using radio links with the Cassini spacecraft. Nature **425**, 374–376 (2003)

29. K. Danzmann, LISA: an ESA cornerstone mission for a gravitational wave observatory. Class. Quantum Grav. **14**, 1399 (1997)
30. P. Amaro-Seoane et al., *Laser Interferometer Space Antenna.* Preprint at http://arxiv.org/abs/1702.00786 (2017)
31. J. Luo, L.S. Chen, H.Z. Duan, Y.-G. Gong, S. Hu, J. Ji, Q. Liu, J. Mei, V. Milyukov, M. Sazhin et al., TianQin: a space-borne gravitational wave detector. Class. Quantum Grav. **33**, 035010 (2016)

Chapter 2
A Brief Introduction to the TianQin Project

Jun Luo

Abstract TianQin is a space-based gravitational wave detector expected to launch around 2035. It will detect gravitational waves in the frequency band 0.1 mHz $\sim$ 1 Hz, offering the chance to detect gravitational waves from a large variety of sources, such as massive black holes, stellar mass black holes, Galactic compact binaries, and so on. The mission was officially proposed in 2014 and has passed a couple of major milestones.

2.1 Introduction

Our team started research on gravitational experiments in the early 1980s. In 2000, we began to research on two core technologies of space-based gravitational wave (GW) detection: inertial sensing and inter-satellite laser interferometry. In 2013, our first electrostatic inertial sensor was successfully tested in orbit, marking a historical moment when we had the necessary technology basis to start thinking about real space-based GW detection missions. The TianQin project was then officially proposed during an international workshop on space-based gravitational experiments in 2014.

A special feature of TianQin is its geocentric orbit. To achieve a stable constellation, the optimal orbital radius for TianQin is about 100,000 km. Due to their relative proximity to Earth compared to those in heliocentric missions, the TianQin spacecraft are expected to be easier to deploy, measure and communicate. But the use of geocentric orbit also brings unique challenges, which I will explain in more detail later. We expect to launch TianQin around 2035.

TianQin is not the only space-based GW detector planned for the 2030s. There are also the ESA led LISA [1, 6], the Taiji project proposed by the Chinese Academy of Sciences [7], and also DECIGO from Japan [9]. TianQin and LISA, being two

---

J. Luo (✉)
MOE Key Laboratory of TianQin Mission, TianQin Research Center for Gravitational Physics, School of Physics and Astronomy, Frontiers Science Center for TianQin, Gravitational Wave Research Center of CNSA, Sun Yat-sen University (Zhuhai Campus), Zhuhai, China
e-mail: junluo@mail.sysu.edu.cn

J. Luo et al. (eds.), *Proceedings of the 7th International Workshop on the TianQin Science Mission; 25–26 April 2024, Hong Kong SAR, China*, Springer Proceedings in Physics 428, https://doi.org/10.1007/978-981-96-7803-7_2

representative missions using geocentric and heliocentric orbits, respectively, are notably different in terms of their constellation motion and armlength, and hence their response to GWs. In Fig. 2.1, the sky averaged sensitivity curve of TianQin is compared to those of LISA and all the current ground-based detectors.

In the following, I briefly introduce the science, the concept challenges, and the current progress with the TianQin project.

2.2 Science with TianQin

By detecting GWs in the milli-hertz (0.1 mHz $\sim$ 1 Hz) range, TianQin is sensitive to a variety of GW sources, including:

- The revolution of galactic compact binaries;
- The inspiral of stellar mass black hole binaries;
- The inspiral and plunge of binary systems with extreme/large mass ratios;
- The inspiral-merger-ringdown of massive black hole binaries;
- Stochastic GW background, including those from unresolved astrophysical sources and from possible energetic processes in the early universe.

Different types of sources can be detected to different distance, hence different time and redshifts, in the universe. Starting from the nearby galactic compact binaries, TianQin can detect stellar mass black hole binaries up to redshift $z \sim \mathcal{O}(0.1)$, extreme/large mass ratios up to redshift $z \sim \mathcal{O}(1)$, massive black hole binaries up to

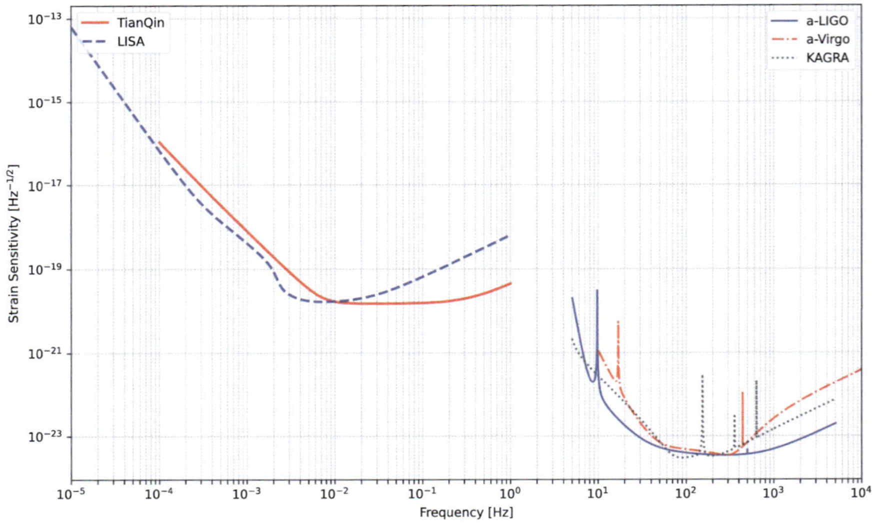

Fig. 2.1 The sky averaged sensitivity of TianQin [10], together with those of LISA [2], Advanced LIGO [3], Advanced Virgo [4] and KAGRA [16] as comparison

$z \sim \mathcal{O}(10)$, and potentially stochastic GW background from the first order phase transition of the universe within picoseconds after its birth.

By probing different epochs of the universe, TianQin can make huge impact on astrophysics, fundamental physics and cosmology. Some major scientific goals of TianQin include:

Astrophysics Discovering massive and intermediate mass binary black hole systems and compact binary systems in globular clusters; revealing the structure and material distribution of the Milky Way, the central environment of galaxies surrounding massive black holes, the formation and evolution mechanisms of compact binary systems;

Fundamental physics High precision test of general relativity in the strong field regime;

Cosmology Accurately characterizing the evolution history of the universe, and measuring dark matter and dark energy related parameters in a wider spacetime region;

Exploring new physics Revealing the dynamics of symmetry breaking and first-order phase transition in the early universe.

A summary of the scientific potential of TianQin can be found in [15]. A systematic study on the potential of joining two space-based detectors such as TianQin and LISA can be found in [19].

2.3 Concept Challenges

The choice of geocentric orbit brings some advantages for TianQin. For example, due to its proximity to Earth, TianQin has: (1) a relatively short transfer time at the order of a couple of months during deployment; (2) relatively stable orbit and constellation during science observations, with the variation of armlength being less than 1% and the relative velocities among satellites below 5 m/s; (3) relatively high orbit determination precision, at the order of a few centimeters; (4) relatively short communication time and high data transmission rate with the ground segment [24]; and so on.

TianQin also faces some challenges. There are environmental disturbances that are common for all space-based GW missions, such as space electric and magnetic fields, cosmic rays, and so on. But due to the particular orbital configuration of TianQin, especially its vertical orbital plane with respect to the ecliptic, the gravity field, sun radiation, thermal and plasma environment faced by TianQin is more complicated than heliocentric missions such as LISA. All these bring challenges and has caused concern about the feasibility of the TianQin mission concept.

The first question is about the orbit stability of TianQin. Before TianQin, no one had tried in GW detection missions to use a constellation whose orbital plane is vertical to the ecliptic. The presumption had been that, no stable orbit can be found in such a configuration. Through simulation, we find that the number of stable orbits for TianQin is unlimited, as long as the orbital radii are properly chosen and the orbit planes are nearly vertical to the lunar orbit [22, 23, 28].

The second question is about the effect of irregular gravity field in the Earth-Moon system. The TianQin spacecraft are facing complicated gravity environment due to their proximity to the Earth. The Earth itself is not a perfect sphere and there is significant asymmetric distribution of mass on Earth, which leads to irregular distribution of the Earth gravity field. There is also the Moon, which is also a significant source of gravitational attraction. Could such irregularities in the near Earth gravity field cause too much disturbance on the TianQin spacecraft so as to significantly affect the GW detection? With numerical simulation, we find that the near Earth gravity field can indeed cause significant acceleration noise on the test masses. Fortunately, the acceleration noise induced by the Earth gravity field decreases steeply around 10^{-4} Hz, and it is below the target sensitivity frequency band of TianQin [12, 26].

The third question is about Moon eclipse. The orbital radius of the TianQin spacecraft is smaller than that of the Moon. The moon can get in between the Sun and the Earth every month, and so there is chance for Moon eclipse on the TianQin spacecraft, causing dramatic variations in the solar radiation received by the satellites, and the resulted thermal shock could cause interruptions to GW detection. We find that, by using Moon resonant orbits, it is possible to find orbits that are not only stable, but also totally free of Moon eclipse during the science observation periods [21].

The fourth, which is also the earliest known, problem faced by TianQin is the solar radiation. Due to the use of geocentric orbits, the orbital plane of TianQin is nearly fixed in space. So the Sun will revolve around the constellation and pass through the orbital plane of TianQin once every half year. If the Sun light gets into the telescopes, it will likely cause too much disturbance and can interrupt GW detection. To deal with this problem, we have come up with the "3 month on + 3 month off" detection scheme for TianQin [13]. The orbital plane of TianQin is currently set to directly face the ultra-compact binary source, RX J0806.3+1527 [17], which is only about 4.7° below the ecliptic. From the beginning of June to the beginning of September, and from the beginning of December to the beginning of March, the Sun light is at large angles with the TianQin orbital plane (and also the Sun shield on the TianQin spacecraft). There is minimal disturbance from the Sun during these periods and TianQin will do scientific observations. During other times of the year, the Sun will come close to the orbital plane of TianQin and the he detector will be put into a safe mode.

The large variation of the solar angle also means there is large variation in the heat load on the TianQin spacecraft. But within the TianQin sensitivity frequency band, the radiation noise is dominated by the solar constant fluctuation, and which is the same as for other space-based detectors. There is also radiation noise from the Earth but which can be neglected [5].

Another important challenge for TianQin is the space plasma. Being close to the Earth means that TianQin is close to regions with high density of space plasma. Due to the long distance in inter-satellite laser propagation, the space plasma could become a significant source of phase noise. Our preliminary study show that plasma induced phase noise is subdominant except during strong magnetic storms. We expect more study to minimize the potential impact of the space plasma [8, 11, 18].

To sum up, we find that there are some unique challenges to the TianQin mission concept, but none is challenging enough to fundamentally limit the mission from achieving its scientific goals. In fact, the most important challenges are still the technologies that are needed, which I will discuss in the next section.

2.4 Technology Progress

To successfully detect GWs in space, TianQin will need the following [13]:

- Inter-satellite laser interferometry. To measure the effect of GW signals, and the noise in the displacement measurement by inter-satellite laser interferometry should be lowered to the order 10^{-12} m/Hz$^{1/2}$;
- Inertial reference. The residue acceleration noise on the test masses, which serve as the "reflecting mirrors" like in a Michelson interferometer, should be lowered to the order 10^{-15}m/s^2/Hz$^{1/2}$;
- Ultra-stable spacecraft and constellation. These will provide the necessary quiet and stable environment needed for the key instruments to function as designed.

To bring the required technology to maturity, we have adopted the so called "0123" technology roadmap:

Step "0" is to acquire the capability of lunar laser ranging, with which one can obtain high-precision orbit information for the TianQin spacecraft;
Step "1" is a single satellite mission, which is to use a single satellite to test the inertial reference technology;
Step "2" is a duo satellite mission, which is to use a pair of satellites to test the inter-satellite laser interferometry technology;
Step "3" is a triple satellite mission, which is to launch a constellation of three satellites to form the space-based GW detector, TianQin, and to carry out GW detection missions in space.

Several major milestones have been reached along this roadmap.

In step "0", we have acquired the lunar laser ranging (LLR) capability. The TianQin spacecraft are using geocentric orbits, so it is possible to use laser ranging technology to obtain high precision orbit information. In order to prepare for the task, we start by doing LLR. We built the TianQin laser ranging station in the beginning of 2019 and get the first LLR signal in June 2019. A preliminary analysis shows that the statistical errors in most of the normal points are better than 1 cm, which is enough for the precision orbit determination of the TianQin spacecraft [25].

In step "1", we have used the TianQin-1 experimental satellite to test inertial reference technology in space. In more detail, the satellite is designed to test six technologies in space: Inertial sensing; Micro-Newton thruster; Drag-free control;

Laser interferometry; Stable temperature control; Stable center-of-mass control. The satellite uses a helio-synchronous orbit with an altitude of about 600 km. The mass of the satellite is about 100 kg and the mission lifetime is designed to be 6 months. The satellite eventually survived 13 months in space. All the experiments on the satellite have been successful performed. As a most important result from the satellite, inertial sensors with sensitivity at the order 10^{-12} m/s^2/Hz$^{1/2}$ has now been independently verified [14, 20, 27].

Our next big mission is the TianQin-2 experimental satellite mission for step "2". The main objective of the mission is to use a pair of satellites that are about 200 km apart to verify the following key technologies: narrow line-width laser; frequency stabilization; ultra-stable optical bench; weak light phase locking; inter-satellite beam pointing; satellite-satellite pointing control. The mission has been officially approved by CNSA and is expected to launch around 2026.

In the meantime, we are also carrying out systematic study on the step "3" GW detection mission. Some of the current research focus are: mission study, inertial sensor, laser interferometry, optical and mechanical integration, satellite bus, drag-free control, inter-satellite ranging, communication and clock comparison, orbit determination, navigation and control, launch. We hope the mission can get official approval before 2030.

Acknowledgements This work has been supported in part by the Guangdong Major Project of Basic and Applied Basic Research (Grant No. 2019B030302001).

References

1. P. Amaro-Seoane, et al. Laser Interferometer Space Antenna
2. S. Babak, A. Petiteau, M. Hewitson, LISA Sensitivity and SNR Calculations
3. L. Barsotti, S. Gras, M. Evans, P. Fritschel, The updated Advanced LIGO design curve (2018)
4. C. Berry, S. Fairhurst, B. OReilly, M. Razzano, P.J. Sutton, Prospects for Observing and Localizing Gravitational-Wave Transients with Advanced LIGO, Advanced Virgo and KAGRA (2019)
5. H. Chen, et al. Thermal environment analysis for TianQin. Class. Quant. Grav. **38**(15), 155015 (2021)
6. M. Colpi, et al. LISA Definition Study Report
7. W.-R. Hu, Y.-L. Wu, The Taiji program in space for gravitational wave physics and the nature of gravity. Natl. Sci. Rev. **4**(5), 685–686 (2017)
8. Y.-D. Jing et al., Plasma noise in TianQin time-delay interferometry. Phys. Rev. D **106**, 082006 (2022)
9. S. Kawamura et al., Current status of space gravitational wave antenna DECIGO and B-DECIGO. PTEP **2021**(5), 05A105 (2021)
10. E.-K. Li, et al., GWSpace: a multi-mission science data simulator for space-based gravitational wave detection
11. L.-F. Lu et al., Effects of the space plasma density oscillation on the interspacecraft laser ranging for TianQin gravitational wave observatory. J. Geophys. Res.: Space Phys. **126**(2), e2020JA028579 (2021)
12. C. Luo, X. Zhang, Effect of Earth-Moon's gravity on TianQin's range acceleration noise. II. Impact of orbit selection. Phys. Rev. D **105**, 102007 (2022)

13. J. Luo et al., TianQin: a space-borne gravitational wave detector. Class. Quant. Grav. **33**(3), 035010 (2016)
14. J. Luo et al., The first round result from the TianQin-1 satellite. Class. Quant. Grav. **37**(18), 185013 (2020)
15. J. Mei et al., The TianQin project: current progress on science and technology. PTEP **2021**(5), 05A107 (2021)
16. Y. Michimura, et al., Estimated sensitivity for auxiliary degrees of freedom of the KAGRA interferometer (2020)
17. G.H.A. Roelofs, A. Rau, T.R. Marsh, D. Steeghs, P.J. Groot, G. Nelemans, Spectroscopic evidence for a 5.4-minute orbital period in HM Cancri. Astrophys. J. Lett. **711**, L138 (2010)
18. W. Su, et al., Analyses of laser propagation noises for TianQin gravitational wave observatory based on the global magnetosphere MHD simulations. Astrophys. J. **914**
19. A. Torres-Orjuela, S.-J. Huang, Z.-C. Liang, S. Liu, H.-T. Wang, C.-Q. YeY.-M. Hu, J. Mei, Detection of astrophysical gravitational wave sources by TianQin and LISA
20. C. Xiao et al., Drag-free control design and in-orbit validation of TianQin-1 satellite. Class. Quant. Grav. **39**(15), 155001 (2022)
21. B. Ye et al., Eclipse avoidance in TianQin orbit selection. Phys. Rev. D **103**, 042007 (2021)
22. B. Ye, X. Zhang, Effects of lunisolar perturbations on TianQin constellation: An analytical model. Phys. Rev. D **109**, 083033 (2024)
23. B.-B. Ye et al., Optimizing orbits for TianQin. Int. J. Mod. Phys. D **28**(09), 1950121 (2019)
24. Z. Yi et al., Data transmission analysis and communication scheme design for TianQin mission. IEEE Access **12**, 42585–42593 (2024)
25. C. Zhang et al., The facilities and performance of TianQin laser ranging station. Class. Quant. Grav. **39**(12), 125005 (2022)
26. X. Zhang et al., Effect of Earth-Moon's gravity on TianQin's range acceleration noise. Phys. Rev. D **103**, 062001 (2021)
27. A.-N. Zhou, L. Cai, C.-Y. Xiao, D.-Y. Tan, H.-Y. Li, Y.-Z. Bai, Z.-B. Zhou, J. Luo, Non-gravitational force measurement and correction by a precision inertial sensor of TianQin-1 satellite. Class. Quant. Grav. **39**(11), 115005 (2022)
28. M.-Y. Zhou et al., Orbital effects on time delay interferometry for TianQin. Phys. Rev. D **103**, 103026 (2021)

Part II
Science and Data Analysis with TianQin

Chapter 3
Gravitational Wave Astrophysics with TianQin: A Brief Progress Review

Yi-Ming Hu

Abstract As a space-borne gravitational wave observatory, TianQin can observe a large variety of gravitational wave sources. The rich signals can be composed by different types of astronomical systems, like Galactic compact binaries, inspiral of stellar mass black holes, merger of massive black holes, and extreme mass ratio inspirals. The incoherent summation of these signals can also form a stochastic gravitational wave background. Using the future TianQin observation, it is possible to put stringent constraints on fundamental science, like the expansion history of our universe. In this work, we provide a brief review of these topics that contributes to the 7th International Workshop on the TianQin Science Mission.

3.1 The TianQin Observatory

TianQin is a proposed space-borne gravitational wave mission. Three satellites are expected to be launched in the 2030s to form the constellation. The satellites will protect the test masses in the core, shielding them from external disturbances, so that the test masses can keep in a free-falling status and closely follow the geodesics. High-precision laser interferometry will be used to measure the small changes in the length between test masses, which reflects the key information of the sources [1]. According to the design, the TianQin can detect gravitational waves in the milli-Hertz frequency band, covering a wide range of sources coming from a rich variety of locations, from as close as our cosmic backyard, to as far as the end of the observable universe [2–6].

The sensitivity of the TianQin can be indicated by the Power Spectral Density (PSD), which characterizes the frequency-dependent fluctuations of the noise. For

Y.-M. Hu (✉)
MOE Key Laboratory of TianQin Mission, TianQin Research Center for Gravitational Physics, Sun Yat-sen University (Zhuhai Campus), Zhuhai, China
e-mail: huyiming@mail.sysu.edu.cn
URL: https://yiminghu-sysu.github.io/

School of Physics and Astronomy, Frontiers Science Center for TianQin, CNSA Research Center for Gravitational Waves, Sun Yat-sen University (Zhuhai Campus), Zhuhai, China

J. Luo et al. (eds.), *Proceedings of the 7th International Workshop on the TianQin Science Mission; 25–26 April 2024, Hong Kong SAR, China*, Springer Proceedings in Physics 428, https://doi.org/10.1007/978-981-96-7803-7_3

the convenience of calculation, the sky-averaged response is often absorbed in the expression of the sensitivity curve. The effective PSD of the TianQin $S_n(f)$ can be approximated as [2]

$$S_n(f) = \frac{1}{L^2}\left[\frac{4S_a}{(2\pi f)^4}\left(1+\frac{10^{-4}\text{Hz}}{f}\right)+S_x\right]\times\left[1+0.6\left(\frac{f}{f_*}\right)^2\right] \tag{3.1}$$

Here, $L=\sqrt{3}\times10^8$m is the arm-length between any pair of satellites, $S_a = 10^{-30}\text{m}^2\,\text{s}^{-4}\,\text{Hz}^{-1}$ is the acceleration noise, $S_x = 10^{-24}\text{m}^2\,\text{Hz}^{-1}$ is the displacement measurement noise, and $f_* = c/2\pi L = 0.28$Hz is the transfer frequency. We can observe that the PSD can be determined by the three parameters, the acceleration noise determines the low-frequency behavior, the displacement measurement noise determines the level of the bucket, and the armlength/transfer frequency determines the high-frequency sensitivity.

The TianQin satellites are designed to follow a geocentric orbit, with the orbit height of 10^8m. The orbital period around the Earth is around 3.64 d, which means that the TianQin satellites will fly around 100 cycles annually. The choice of the geocentric orbit also means that the orbital plane of the TianQin satellites is also fixed. It is proposed that the norm of the TianQin plane is the white dwarf binary system RXJ0806.3+1527 (also known as HM Cancri, hereafter J0806). The Electromagnetic (EM) observation of J0806 revealed its orbital frequency as well as its sky location, therefore, TianQin can use the binary system as a verification binary to calibrate the instrument performance. J0806 has an ecliptic coordinates as $\lambda = 120.4°$, $\beta = -4.7°$. Being close to the ecliptic planes has extra advantages, since the orbital, as well as thermal stabilities, are both close to ideal under this choice [7–9] However, when the direction of the Sun is too close to the orbital plane, the thermal shielding will be no longer ideal, and the detector will switch off the science mode for better protection. This led to the "three months on + three months off" work scheme of the TianQin mission, and the nominal duty cycle is 50% [1, 10].

With the operation of TianQin, sources within 0.1 mHz 1 Hz can be recorded by TianQin, providing a powerful tool to decipher the dark universe. Theoretical work indicates that TianQin can observe sources near and far, light and heavy. In the Milky Way, it is expected that there exist hundreds of millions of double white dwarfs and thousands of them can be detected by TianQin [2]. Ground-based gravitational wave detectors have observed the late inspiral and merger of the other type of stellar-mass compact binaries, namely the stellar-mass binary black holes. TianQin can observe the early inspirals emitted years before the merger [3]. If massive black holes capture stellar-mass black holes, the heavily uneven binary will form the Extreme Mass Ratio Inspiral (EMRI), where the small black hole follows the geodesic and delivers rich information about the central black hole through complicated orbits [5]. Binaries of massive black holes can emit strong Gravitational Wave(GW) signals so that if the masses are ideal, even if they are on the edge of the observable universe, TianQin will be able to make reliable detections [4]. In addition to the individual sources, the superposition of various GWs can form a noise-like stochastic background [6],

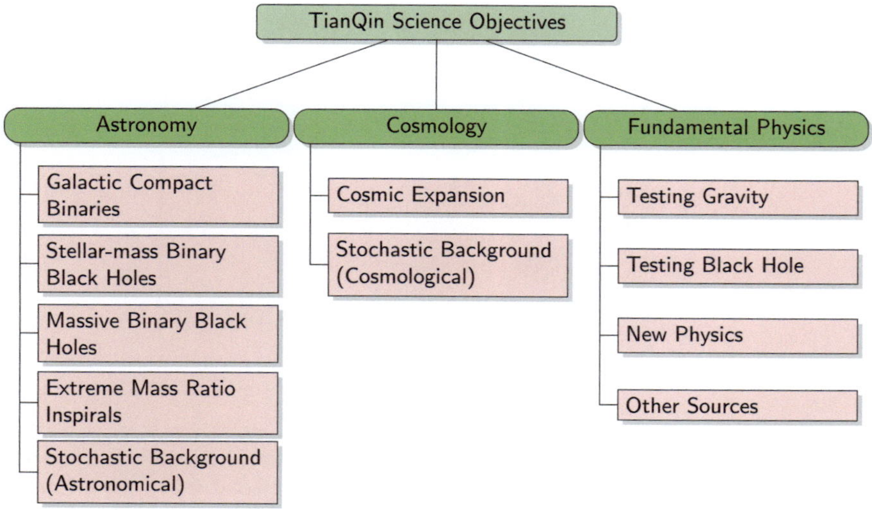

Fig. 3.1 Science objectives of the TianQin observatory

and their original can be either astrophysical or cosmological. TianQin's observation of these systems will not only provide an important message about the sources, but also provide ideal laboratories to study the nature of gravity, the black hole, and the expansion of the universe. It might also reveal pivotal evidence for new physics beyond the standard model [10]. We summarise the major scientific objectives in Fig. 3.1.

In this document, we aim to provide a brief overview of GW astronomy with TianQin. We will focus on what sources can TianQin observe, the expected detection rate, and anticipated parameter estimation precision. Each of the following sections will be dedicated to stellar-mass compact binaries, extreme mass ratio inspirals, massive black hole binaries, and stochastic gravitational wave backgrounds. At the end, we will briefly summarize the GW astronomy with TianQin and the progress of the project on GW astronomy research.

3.2 Gravitational Wave Sources

3.2.1 Stellar-Mass Compact Binaries

3.2.1.1 Galactic Compact Binaries

Observation indicates that half of the stars are in binary systems. When the stars run out of fuel, the gravitational pull will force the stars to crash into compact objects like white dwarfs, neutron stars, or black holes, the exact form depending on the

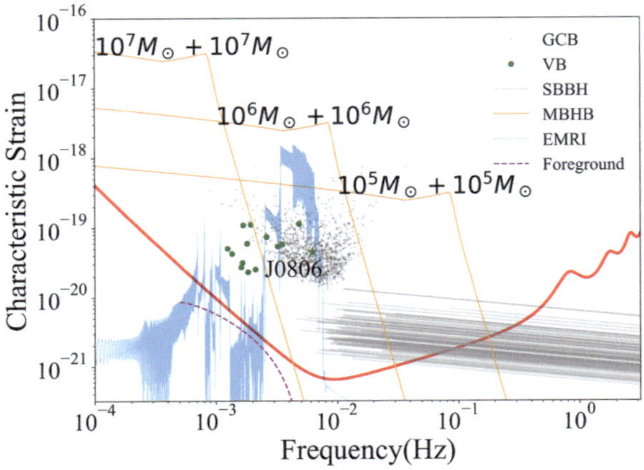

Fig. 3.2 Characteristic strain of typical sources. Gray dots represent observed white dwarf binaries through optical telescopes, while green dots indicate the strongest predicted white Galactic binaries. Gray/yellow lines indicate the inspiral (and merger) of stell stellar-mass/massive binary black holes. The blue line represents a typical extreme mass ratio inspiral, and the purple dashed line shows the foreground from the Galactic compact binaries

remanent mass. Since the mass distribution of the stars is heavily tail-dominated, the vast majority of the dead stars exist in the form of white dwarfs, and those in binary systems will form double white dwarfs. During the evolution, processes like mass transfer, common envelope phase, and/or dynamical encounter will "harden" the binaries, namely increasing their gravitational binding energy. Even though only a small fraction of the compact binaries in the Galaxy is expected to be so closely bound that they emit GW signals in the mHz frequency band, the sheer volume of the total compact binaries makes the mHz GW window packed with Galactic compact binaries (Fig. 3.2).

In principle, one can assess the GW properties of the Galactic compact binaries through either theoretical modeling or astronomical observations. However, the identification and confirmation of the binarity for Galactic compact binaries is challenging. Attempts have been made to make systematic searches for double white dwarfs using data from equipment like Gaia and the Zwicky transient facility. Still, there are only dozens of Verification Binary (VBs) for TianQin, which are systems identified as double white dwarfs, and the expected Signal-to-Noise Ratio (SNR) larger than 5 [11]. The scarcity of VBs only implies the difficulty of observation through optical observation. Using population synthesis methods, theoretical modeling of the Galactic compact binaries predicts that with a five-year mission lifetime, TianQin is expected to observe about 9,000 double white dwarfs, enriching the sample size by two orders of magnitudes and can provide precious clues for the complex processes during the late time binary evolution [2].

The GW observation from TianQin can not only contribute to observing thousands of double white dwarfs, but also provide highly precise measurement of the physical parameters. For example, for most of the sources, the orbital periods and amplitudes can be measured to the precision of 10^{-7} and 20%, respectively. About 40% of the detectable sources can be localized to within 1 deg^2 [2].

Since the Galactic compact binaries can produce multi-messenger emissions, they are ideal testbeds for independent measurement of Galactic structure [12], testing the theory of gravity [13], or even the discovery of exoplanets [14].

3.2.1.2 Stellar-Mass Binary Black Holes

Since the operation of the advanced LIGO in 2015, nearly a hundred GW events have been officially reported, with more being accumulating [15]. The majority of these events are mergers of stellar-mass binary black holes. For the ground-based detectors, these black hole mergers typically last only seconds in duration. However, in the mHz frequency band, the evolution of the binary is much slower, so space-borne GW observatories like TianQin can observe the early inspiral of the stellar-mass binary black holes [3]. In principle, TianQin can identify the GW signals years or even decades before the final merger, making room for well-prepared EM follow-ups.

Shortly after the detection of the first GW event GW150914, optimistic predictions were made that space-borne GW missions can observe up to thousands of stellar-mass binary black holes [16]. During the past decade, we have accumulated more detections to form a more comprehensive understanding of the population properties of the stellar-mass binary black holes, with careful calculations the detection rate has been updated to a lower value, that TianQin can observe about a dozen such inspirals during the five year mission lifetime [3]. This means that space-borne GW detectors can observe only nearby stellar-mass binary black holes, roughly corresponding to the redshift limit of 0.2.

Although the expected detection rate is not high, TianQin is still expected to make very precise measurements for the physical parameters. For instance, even though the binaries evolve off the sensitive band long before the merger, TianQin can still predict the final merger time to the precision of the sub-second level. The sky location can also be pinpointed to the level of 1 square degree, which is comparable to or even smaller than many of the field of view for many survey telescopes. The high precision in spatial and temporal parameters makes the EM follow-up easy to implement. Other parameters, like the orbital eccentricity, can also be determined to the level of 10^{-4}. At first glance, the high precision might seem counter-intuitive, since the ground-based detectors are expected to measure the mergers with higher SNRs, yet the current detector network can only constrain the sky location to the level of tens of square degrees. The fundamental reason lies in the fact that TianQin can accumulate signals over a long period of time. Over the years of operation, the modulation due to Doppler shift and antenna pattern change has made TianQin very sensitive to tiny changes in the parameter space [3].

The high precision is rewarding. For a typical source, the spatial error volume is so small that on average there is only one galaxy within, making it highly likely to identify the host galaxy for the binary black holes. This means that even in the lack of any EM counterpart, one can still rely on TianQin's GW observation to draw meaningful conclusions on questions like the expansion rate of the universe [17]. The prospect of multi-band GW observation also implies the possibility of constraining the environment [18], the theory of gravity [19], or their formation mechanisms [20].

3.2.2 Extreme Mass Ratio Inspirals

Observations have revealed the ubiquitous existence of massive black holes within the center of galaxies, which are often crowded with stellar-mass compact objects. If a stellar-mass black hole gets perturbed and becomes very close to the central massive black hole, they will form a highly unequal binary called EMRI. The smaller black hole follows the geodesic, and the non-linear effect predicted by the general theory of relativity makes the orbit highly complicated. Through its long time GW observation, it is possible to use EMRI signals to put stringent constraints on the theory of gravity, since any tiny deviation will manifest itself over the long-term phase shift accumulation, while the GW observation is expected to measure the phase parameters to extremely high precision. Therefore, EMRIs are expected to be ideal laboratories for extreme gravity.

The quantitative prediction on the EMRI detection rate is quite uncertain, this is because different conditions can lead to vastly different event rates, while current observations have very weak constraints on these conditions. That being said, attempts have been made to quantitatively predict TianQin's detection rate. EMRIs systems within redshift of 2 are possible to be detected with TianQin. Studies indicate that unless in the extremely pessimistic scenario, TianQin is expected to make at least a number of EMRI detections. In the most optimistic case, the detection rate for TianQin can be as high as $\sim$400 per year.

Similar to the stellar-mass binary black holes, the TianQin can constrain the physical parameters to a very precise level. For example, the mass parameters, the spin, and the eccentricity can all be constrained to the relative precision of 10^{-6}. In comparison, extrinsic parameters that only affect the amplitude are less well-constrained, for example, the sky localization ability can only be limited to 10 deg^2, and the luminosity distance can only be constrained to 10%.

The excellent precisions of intronic parameters make EMRI systems one of the most ideal laboratories to examine gravity theories in detail. For example, one can use TianQin's observation of EMRI to test the no-hair theorem [21, 22], or to constrain the compact objects' tidal deformability [23]. The future observation of TinQin can help provide key information on EMRI formation, which in turn could greatly improve the precision of GW cosmology, constraining the Hubble constant to a sub-percent level [24].

3.2.3 Massive Black Hole Binaries

The existence of massive black holes in the center of galaxies has been widely accepted, and the imaging of the black hole shallow has been made possible with the Event Horizon Telescope [25, 26]. The mergers of galaxies will bring the central massive black holes together, which eventually lead to the the merger of the binary massive black holes. Although detailed understanding has been under debate long ago [27], the evidence that a stochastic background in the nanohertz gravitational wave window shows strong support for the ubiquitous existence of merging binary massive black holes [28–31]. In addition to the nanohertz signal, these systems can also emit strong GW in the millihertz frequency band, especially for the merger phase as well as for the relatively lighter systems.

The heavier a binary system is, the stronger the GW it emits. For binaries of massive black holes, the strength of the GW signal is so strong that even if a pair of massive black holes merge at the edge of the observable universe, as long as the frequency hits the sweet spot of TianQin's sensitivity band, it will still be observable with the SNR larger than 10. Assuming different astronomical models for the merger history as well as the growth mechanism of massive black holes, we predict that the realistic detection rate ranges from a few to a few dozen per year [32]. If TianQin collaborates with other space-borne GW detectors like LISA, the detection rate can be further improved, especially the sky localization precision can be drastically improved [33].

The largest uncertainty in the rate prediction lies in the formation mechanism of massive black holes, also known as the "seeding" mechanism. Traditional stellar evolution models predict that massive black holes seed from the remnant of population III stars, which is at most hundreds of solar masses. However, this "light seed model" encounters difficulty explaining the observed massive black holes in high redshift, which means that the black holes have to go through an extremely efficient growth stage, which could exceed the Eddington limit. The "heavy seed models" are proposed to resolve this tension by introducing mechanisms like direct collapses, so that seed black holes can be born with high masses [34]. In order to verify and distinguish the seeding mechanism, one needs to depict the mergers at high redshift with high precision, and TianQin can do exactly that: for mergers of massive black holes happen at redshift 15, TianQin can pinpoint both the mass parameters and the distance to the level of 10% [32].

Massive black hole mergers can also be ideal targets for performing multi-messenger observations. The merger of massive black holes can trigger instabilities in the accretion disks, potentially causing EM flares. The coordinated observation of both GW and EM signals can uncover critical links between the massive black holes and their host galaxies, the birth and growth of massive black holes, etc. Notice that the SNR accumulates slowly in the inspiral stage, and changes dramatically during the final merger stage. As SNR increases, the localization error quickly shrinks [35]. Therefore, successful multi-messenger observation relies heavily on the ability to transmit and analyze real-time data. Thanks to the geocentric orbit configuration,

TianQin has the unique advantage of stable and low-latency data downlink, facilitated with a real-time data analysis pipeline, it is possible to coordinate observation plans with EM telescope [36].

3.2.4 Stochastic Gravitational Wave Background

Those GW signals that have large enough SNR will be resolved and analyzed individually. However, for weak signals that can not be individually resolved, their non-coherent superposition will form a stochastic background. Even if the resolvable sources will not be matched and removed perfectly, the residual will form a foreground, especially the Galactic foreground can be comparable to the noise PSD. The statistical properties of the Stochastic Gravitational Wave Background (SGWB) can be indistinguishable from the instrumental noise. Therefore, if one has no *a* priori information on the instrumental noise PSD, it is impossible to detect SGWB from a single detector. In the literature of SGWB observation based on LISA, the *n*ull channel method is proposed. It utilized the fact that a triangle detector has three readouts, while the GW only has two polarizations. The other method is to use multiple independent detectors and cross-correlate their data. The SGWB from different detectors will have non-zero correlations, while the noise is expected to be not correlated.

The origin of SGWB can be either astrophysical or cosmological. In the millihertz band, the astrophysical SGWB is composed of all unresolvable sources mentioned above, while cosmological SGWB can include first-order electroweak phase transition, primordial GW, inflation, and cosmic defects. We limit the scope of this document to the astrophysical origin SGWB, and leave the discussion of the cosmological SGWB for future studies. Since the majority of the unresolvable Galactic binaries are crowded around 1mHz, and TianQin's sensitivity is optimal for slightly higher frequencies, this means that the foreground is a less severe issue for TianQin compared with LISA. After a five-year mission lifetime, the strongest Galactic compact binaries will be identified and removed, and the remaining will be lower than the noise PSD. That being said, using either the null channel or the cross-correlation method, TianQin can observe strong foreground, and the SNR for SGWB from stellar-mass binary black holes can reach 10.

If one relies on the null channel method, then the uncertainty of the noise PSD has to be very small, which might not be realistic [37]. By joining the game, TianQin enables the possibility of cross-correlation, opening the window for reliable analysis of the SGWB. TianQin will also be a game changer in terms of the SGWB anisotropy analysis. For certain multipole components, the TianQin-LISA network will have several orders of magnitude improvement over a single detector.

3.3 Summary and Discussion

In summary, TianQin can observe a rich range of GW signals, ranging from the local Galactic compact binaries, the nearby stellar-mass binary black holes, the extreme mass ratio inspirals to an intermediate distance, and the massive black hole binary mergers to the edge of the observable universe. In addition to the individually resolvable sources, TianQin can also detect their superposition as a stochastic gravitational wave background. The detection rate can be highly uncertain and can differ by orders of magnitude, but it's highly likely that TianQin will observe a large number of sources, probably dominated by Galactic compact binaries. Using these TianQin detections, one can answer some of the puzzling questions in fundamental physics and astronomies, like constraining the formation channels of the GW sources, exploring their immediate environment, testing the general theory of relativity, probing the expansion history of the universe, etc.

On the basis of the previous studies, a team of astronomers has worked on a TianQin astrophysics whitepaper for two years, which was recently published [38].

The TianQin team does not only focus on the GW astronomy but also pays attention to data analysis skills. So far, a first round of studies has been performed to acquire the simulation ability [39] and to gather basic analysis tools for different sources [33, 36, 40–46] In the future, we plan to make the tool more reliable, more robust, and deal with more realistic issues [47–49].

References

1. J. Luo et al., TianQin: a space-borne gravitational wave detector. Class. Quant. Grav. **33**(3), 035010 (2016)
2. S.-J. Huang, H. Yi-Ming, V. Korol, P.-C. Li, Z.-C. Liang, L. Yang, H.-T. Wang, Yu. Shenghua, J. Mei, Science with the TianQin observatory: preliminary results on galactic double white dwarf binaries. Phys. Rev. D **102**(6), 063021 (2020)
3. S. Liu, H. Yi-Ming, J. Zhang, J. Mei, Science with the TianQin observatory: preliminary results on stellar-mass binary black holes. Phys. Rev. D **101**(10), 103027 (2020)
4. H.-T. Wang, Z. Jiang, A. Sesana, E. Barausse, S.-J. Huang, Y.-F. Wang, W.-F. Feng, Y. Wang, H. Yi-Ming, J. Mei, J. Luo, Science with the TianQin observatory: preliminary results on massive black hole binaries. Phys. Rev. D **100**(4), 043003 (2019)
5. H.-M. Fan, H. Yi-Ming, E. Barausse, A. Sesana, J. Zhang, X. Zhang, T.-G. Zi, J. Mei, Science with the TianQin observatory: preliminary result on extreme-mass-ratio inspirals. Phys. Rev. D **102**(6), 063016 (2020)
6. Z.-C. Liang, H. Yi-Ming, Y. Jiang, J. Cheng, J. Zhang, J. Mei, Science with the TianQin Observatory: preliminary results on stochastic gravitational-wave background. Phys. Rev. D **105**(2), 022001 (2022)
7. X. Zhang, C. Luo, L. Jiao, B. Ye, H. Yuan, L. Cai, G. Defeng, J. Mei, J. Luo, Effect of Earth-Moon's gravity on TianQin's range acceleration noise. Phys. Rev. D **103**(6), 062001 (2021)
8. B. Ye, X. Zhang, Effects of lunisolar perturbations on TianQin constellation: an analytical model. Phys. Rev. D **109**(8), 083033 (2024)
9. H. Chen, C. Ling, X. Zhang, X. Zhao, M. Li, Y. Ding, Thermal environment analysis for TianQin. Class. Quant. Grav. **38**(15), 155015 (2021)

10. J. Mei et al., The TianQin project: current progress on science and technology. PTEP **2021**(5), 05A107 (2021)
11. L. Ren, C. Li, B. Ma, G. Sihao, S.-J. Huang, B. Tang, Y.-m. Hu, A systematic search for short-period close white dwarf binary candidates based on Gaia EDR3 catalog and Zwicky transient facility data. Astrophys. J. Suppl. **264**(2), 39 (2023)
12. V. Korol, E.M. Rossi, E. Barausse, A multimessenger study of the Milky Way's stellar disc and bulge with LISA, Gaia, and LSST. Mon. Not. Roy. Astron. Soc. **483**(4), 5518–5533 (2019)
13. T.B. Littenberg, N. Yunes, Binary white dwarfs as laboratories for extreme gravity with LISA. Class. Quant. Grav. **36**(9), 095017 (2019)
14. N. Tamanini, C. Danielski, The gravitational-wave detection of exoplanets orbiting white dwarf binaries using LISA. Nat. Astron. **3**(9), 858–866 (2019)
15. K. Chatziioannou, T. Dent, M. Fishbach, F. Ohme, M. Pürrer, V. Raymond, J. Veitch, Compact binary coalescences: gravitational-wave astronomy with ground-based detectors. 9 (2024)
16. A. Sesana, Prospects for multiband gravitational-wave astronomy after GW150914. Phys. Rev. Lett. **116**(23), 231102 (2016)
17. L.-G. Zhu, L.-H. Xie, H. Yi-Ming, S. Liu, E.-K. Li, N.R. Napolitano, B.-T. Tang, J. Zhang, J. Mei, Constraining the Hubble constant to a precision of about 1% using multi-band dark standard siren detections. Sci. China Phys. Mech. Astron. **65**(5), 259811 (2022)
18. A. Toubiana et al., Detectable environmental effects in GW190521-like black-hole binaries with LISA. Phys. Rev. Lett. **126**(10), 101105 (2021)
19. A. Toubiana, S. Marsat, E. Barausse, S. Babak, J. Baker, Tests of general relativity with stellar-mass black hole binaries observed by LISA. Phys. Rev. D **101**(10), 104038 (2020)
20. S. Liu, L.-G. Zhu, H. Yi-Ming, J. Zhang, M.-J. Ji, Capability for detection of GW190521-like binary black holes with TianQin. Phys. Rev. D **105**(2), 023019 (2022)
21. T.-G. Zi, J.-D. Zhang, H.-M. Fan, X.-T. Zhang, H. Yi-Ming, C. Shi, J. Mei, Science with the TianQin observatory: preliminary results on testing the no-hair theorem with extreme mass ratio inspirals. Phys. Rev. D **104**(6), 064008 (2021)
22. T. Zi, P.-C. Li, Gravitational waves from extreme-mass-ratio inspirals around a hairy Kerr black hole. Phys. Rev. D **108**(8), 084001 (2023)
23. C.-Q. Ye, J.-H. Chen, J. Zhang, H.-M. Fan, H. Yi-Ming, Observing white dwarf tidal stripping with TianQin gravitational wave observatory. Mon. Not. Roy. Astron. Soc. **527**(2), 2756–2764 (2023)
24. L.-G. Zhu, H.-M. Fan, X. Chen, H. Yi-Ming, J. Zhang, Improving cosmological constraints by inferring the formation channel of extreme-mass-ratio inspirals. Astrophys. J. Suppl. **273**(2), 24 (2024)
25. K. Akiyama et al., First m87 event horizon telescope results. vi. the shadow and mass of the central black hole. Astrophys. J. Lett. **875**(1), L6 (2019)
26. K. Akiyama et al., First sagittarius a* event horizon telescope results. I. The shadow of the supermassive black hole in the center of the Milky Way. Astrophys. J. Lett. **930**(2), L12 (2022)
27. M.C. Begelman, R.D. Blandford, M.J. Rees, Massive black hole binaries in active galactic nuclei. Nature **287**, 307–309 (1980)
28. G. Agazie et al., The NANOGrav 15 yr data set: evidence for a gravitational-wave background. Astrophys. J. Lett. **951**(1), L8 (2023)
29. J. Antoniadis et al., The second data release from the European Pulsar Timing Array - III. Search for gravitational wave signals. Astron. Astrophys. **678**, A50 (2023)
30. D.J. Reardon et al., Search for an isotropic gravitational-wave background with the Parkes Pulsar Timing Array. Astrophys. J. Lett. **951**(1), L6 (2023)
31. X. Heng et al., Searching for the nano-hertz stochastic gravitational wave background with the Chinese Pulsar Timing Array data release I. Res. Astron. Astrophys. **23**(7), 075024 (2023)
32. H.-T. Wang et al., Science with the TianQin observatory: preliminary results on massive black hole binaries. Phys. Rev. D **100**(4), 043003 (2019)
33. J. Gao, Y-M. Hu, E-K. Li, J.-d. Zhang, J. Mei, Bayesian parameter estimation of massive black hole binaries with TianQin-LISA. 1 (2024)

34. F. Pacucci, B. Nguyen, S. Carniani, R. Maiolino, X. Fan, JWST CEERS and JADES active galaxies at z = 4–7 violate the local $M_\bullet$-$M_\star$ relation at $> 3\sigma$: implications for low-mass black holes and seeding models. Astrophys. J. Lett. **957**(1), L3 (2023)
35. W.-F. Feng, H.-T. Wang, H. Xin-Chun, H. Yi-Ming, Y. Wang, Preliminary study on parameter estimation accuracy of supermassive black hole binary inspirals for TianQin. Phys. Rev. D **99**(12), 123002 (2019)
36. H.-Y. Chen, X.-Y. Lyu, E.-K. Li, H. Yi-Ming, Near real-time gravitational wave data analysis of the massive black hole binary with TianQin. Sci. China Phys. Mech. Astron. **67**(7), 279512 (2024)
37. M. Muratore, J. Gair, L. Speri, Impact of the noise knowledge uncertainty for the science exploitation of cosmological and astrophysical stochastic gravitational wave background with LISA. Phys. Rev. D **109**(4), 042001 (2024)
38. E.K. Li, S. Liu, A. Torres-Orjuela, X. Chen, K. Inayoshi, L. Wang, Y.M. Hu, P. Amaro-Seoane, A. Askar, C. Bambi, P.R. Capelo, Gravitational wave astronomy with TianQin. Rep. Prog. Phys. **88**(5):056901
39. E-K. Li et al., GWSpace: a multi-mission science data simulator for space-based gravitational wave detection. 9 (2023)
40. Y. Lu, E-K. Li, Y-M. Hu, J.-d. Zhang, J. Mei, An implementation of galactic white dwarf binary data analysis for MLDC-3.1. Res. Astron. Astrophys. **23**(1), 015022 (2023)
41. X.-T. Zhang, C. Messenger, N. Korsakova, M.L. Chan, Y.-M. Hu, J.-d. Zhang, Detecting gravitational waves from extreme mass ratio inspirals using convolutional neural networks. Phys. Rev. D **105**(12), 123027 (2022)
42. J. Cheng, E.-K. Li, H. Yi-Ming, Z.-C. Liang, J. Zhang, J. Mei, Detecting the stochastic gravitational wave background with the TianQin detector. Phys. Rev. D **106**(12), 124027 (2022)
43. H. Wang, I. Harry, A. Nitz, H. Yi-Ming, Space-based gravitational wave observatories will be able to use eccentricity to unveil stellar-mass binary black hole formation. Phys. Rev. D **109**(6), 063029 (2024)
44. X. Lyu, E.-K. Li, H. Yi-Ming, Parameter estimation of stellar mass binary black holes in the network of TianQin and LISA. Phys. Rev. D **108**(8), 083023 (2023)
45. C.-Q. Ye, H.-M. Fan, A. Torres-Orjuela, J. Zhang, H. Yi-Ming, Identification of gravitational waves from extreme-mass-ratio inspirals. Phys. Rev. D **109**(12), 124034 (2024)
46. X.-T. Zhang, N. Korsakova, M.L. Chan, C. Messenger, Y.-M. Hu, Searching for gravitational waves from stellar-mass binary black holes early inspiral. 6 (2024)
47. Z. Wu, Y.-M. Hu, H.-M. Fan, I.S. Heng, Searching for gravitational-wave bursts with space-borne detectors. Phys. Rev. D **109**(10), 103004 (2024)
48. L. Wang, H.-Y. Chen, X. Lyu, E.-K. Li, Y.-M. Hu, Window and inpainting: dealing with data gaps for TianQin. 5 (2024)
49. Z.-C. Liang, Z.-Y. Li, E.-K. Li, J.-d. Zhang, Y.-M. Hu, Unveiling a multi-component stochastic gravitational-wave background with the TianQin + LISA network. 9 (2024)

Chapter 4
Primordial Intermediate-Mass Binary Black Holes as Targets for Space Laser Interferometers

Konstantin Postnov and Ilya Chekh

Abstract Primordial black holes (PBHs) with log-normal mass spectrum with masses up to $\sim 10^4 - 10^5 M_\odot$ can be created after QCD phase transition in the early Universe at $z \sim 10^{12}$ by modified Affleck-Dine baryogenesis. Using a model binary PBH formation, the expected detection rate of such binary intermediate-mass PBHs by the TianQin space laser interferometer is calculated to be from a few to hundreds events per year for the assumed parameters of the PBH log-normal mass spectrum and abundance consistent with LIGO-Virgo-KAGRA results. Distinctive features of such primordial IMBH mergings are vanishingly small effective spins, possible high redshifts $z > 20$ and lack of association with gas-rich regions or galaxies.

4.1 Intermediate-Mass Black Holes: a Brief Introduction

Black holes (BHs) are topical objects of astrophysical studies. Presently, two broad classes of BHs with different masses are observed by astronomical methods. Stellar-mass black holes ($\sim 3 - 100 M_\odot$) are observed in X-ray binaries and are found by ground-based gravitational-wave interferometers [39]. Supermassive BH (SMBHs, $M \sim 10^5 - 10^9 M_\odot$) are observed in centers of galaxies and active galactic nuclei [31]. The intermediate-mass black holes (IMBHs) with masses $\sim 100 - 10^5 M_\odot$ remain underdetected, likely due to selection effects (see, e.g., [6]). They are possibly found in centers of dwarf galaxies and globular clusters [27, 29, 40] and reported in several LVK events (e.g. GW 190521, GW 190426-190642, 200220-061928, see GWTC-3 catalog [2]. IMBHs can be formed in different ways. They can collapse from very massive Population III stars [44], result from dynamical evolution in dense star clusters [25, 26] or even have a primordial origin [8]. IMBHs in galactic centers can underlie tidal disruption events [21], serve as seeds for SMBH formation [32, 38] and be important sources for space laser gravitational wave interferometers [24].

K. Postnov (✉) · I. Chekh
Sternberg Astronomical Institute, Moscow, Russia
e-mail: k.a.postnov@yandex.ru

J. Luo et al. (eds.), *Proceedings of the 7th International Workshop on the TianQin Science Mission; 25–26 April 2024, Hong Kong SAR, China*, Springer Proceedings in Physics 428, https://doi.org/10.1007/978-981-96-7803-7_4

In this contribution, we address the possibility to detect primordial binary IMBH mergings by a TianQin laser space interferometer [36, 37] promising a rich astronomical output [33]. The specific feature of such events could be their occurrence at high redshifts $z > 20$, before early star formation, where binary IMBHs are not expected to appear by conventional astrophysical mechanisms.

4.2 Primordial Binary Black Holes

Primordial black holes formed in the radiation-dominated era in the early Universe were proposed in the seminal papers [10, 47]. Presently, PBH masses can range from $\sim 10^{15}$ g to supermassive BH values of billion solar masses. Their formation, evolution and observational constraints are subjects of intensive studies (see, for example, reviews [11, 18] and references therein). Binary PBHs became especially interesting after the first LIGO discovery of merging binary stellar-mass black holes [1], which may have a primordial origin [7, 8, 43]. In many subsequent papers, the published results of LIGO-Virgo-KAGRA collaboration [45] were satisfactorily modeled by different groups in terms of binary PBHs [4, 20, 35].

PBHs can comprise significant fraction of cold dark matter [12], be seeds for supermassive black holes in galactic centers [8, 34] and globular clusters [15], generators of cosmic structure [9], and are invoked to solve the present-day questions with early galaxy formation suggested by JWST observations with modest requirements on their abundance [13].

4.2.1 PBHs with Extended Mass Spectrum. Caveat Emptor

The common wisdom is that the PBH can be produced in the early Universe from primordial density perturbations with mass equal to the mass within the cosmological horizon at the time of the PBH formation [10, 47]: at the radiation-dominated stage, $M_h = m_{pl}^2 t$ (in natural units $\hbar = c = k_B = 1$, $G = 1/m_{pl}^2$, m_{pl} is the Planck mass), or $M_h \approx 4.4 \times 10^{38}[g](t/1s)$. The PBH formation from density perturbations is model-dependent (see, e.g., [46]). Here we will concentrate on the particular PBH formation model with extended log-normal mass spectrum suggested in [17, 19]:

$$\frac{dn}{dM} = \mu^2 \exp\left[-\gamma \ln^2\left(\frac{M}{M_0}\right)\right] \tag{4.1}$$

where γ is a dimensionless constant and the parameters μ and M_0 have dimension of mass. In this model, PBHs arise from isocurvature perturbations with high baryonic charge ('high baryon bubbles', HBBs) created before the end of the inflationary stage. These perturbations turned into large density perturbations at the time of the QCD phase transition, $T_{QCD} \sim 100 - 150$ MeV, when massless quarks acquire masses from primordial quark-gluon plasma. The expected mean PBH mass is then $M_0 \simeq 8M_\odot(100\text{MeV}/T_{QCD})^2$ [16]. Note that temperature of the QCD phase transition is still uncertain and depends on the chemical potential (matter density), possible magnetic fields, etc. For example, recent holographic calculations [5] show that T_{QCD} can be appreciably decreased for chemical potential $\mu \sim 0.3$ GeV. Interestingly, this is consistent with expected baryon asymmetry (baryon to photon ratio $\eta = (n_b - n_{\bar{b}})/n_\gamma$) inside HBBs: to create a PBH with mass of the order of $M \sim M_h$, the density contrast should be $\delta\rho/\rho \geq (M_h/M)^2 \sim 1$ at the QCD phase transition, corresponding to $\eta = (\delta\rho/\rho)T_{QCD}/m_q \sim T_{QCD}/m_q \sim 100\text{MeV}/300\text{MeV} \sim 0.3$ (here m_q is the mass of constituent quarks). Thus the decreasing T_{QCD} with non-zero chemical potential helps creating PBHs with higher central mass.

4.2.2 *Intermediate-Mass PBHs*

The extended mass spectrum of PBHs raises the question: how many intermediate-mass PBH can we expect? To substantiate the answer, we will assume the log-normal PBH mass spectrum (4.1). As was shown in [8], the HBBs formed before the end of inflation by the modified Affleck-Dine mechanism [17] could acquire the baryonic charge up to $\sim 10^4 M_\odot$ and produce heavy PBHs after the QCD phase transition. For the log-normal mass distribution (4.1) the generic mass function $\psi(M) = (M/\rho_{pbh})(dn/d\ln M)\,(\int \psi(M)d\ln M = 1)$, where $\rho_{pbh} = f_{pbh}\Omega_{dm}\rho_{cr}$ is the energy density of PBH and $f_{pbh} \leq 1$ is the PBH fraction in dark matter, reads:

$$\psi(M) = \sqrt{\frac{\gamma}{\pi}} e^{-\frac{1}{\gamma}} \left(\frac{M}{M_0}\right)^2 e^{-\gamma \ln^2(M/M_0)}\,. \tag{4.2}$$

Below we shall use masses in units of the central mass M_0, $m \equiv M/M_0$. Motivated by LVK results [4, 20, 35], for numerical estimates below we will use $M_0 = 10, 17, 25 M_\odot$ and $\gamma = 1$. We also will use the standard flat ΛCDM cosmological model with $H_0 = 70$ km/s/Mpc, $\Omega_m = 0.3$ and $\Omega_\Lambda = 0.7$ where necessary. The exponentially small tail of this distribution may seem to imply vanishing fraction of IMBHs. However, integration over the space volumes up to large redshifts available for detection by space interferometers (see [24] and below) and an almost inverse dependence of the primordial binary IMBH merging rate on the comic time $t(z)$ compensate this small fraction and result in a reasonable probability to detect such mergings by TianQin in several years of observations.

To make quantitative estimates, we need to specify the PBH binary formation model. Here we will use the model developed in [41, 42]. This model gives the merging rate of binary PBHs as a function of redshift z

$$\frac{d\mathcal{R}(z)}{d\ln m_1 d\ln m_2} = \frac{1.6\times 10^6}{\mathrm{Gpc}^3\mathrm{yr}} S(m_1, m_2, z) f_{pbh}^{\frac{53}{57}} \left(\frac{t(z)}{t_0}\right)^{-\frac{34}{37}} \psi(m_1)\psi(m_2) M^{-\frac{32}{37}} \eta^{-\frac{34}{37}} \tag{4.3}$$

Here $S \le 1$ is the suppression factor [28, 30, 41, 42] which we will set to 1 below (justified by $f_{pbh} \ll 1$), t_0 is the age of the Universe, $M = (m_1 + m_2)/M_\odot$, $\eta = m_1 m_2/(m_1 + m_2)^2$. It is convenient to change variables $m_1, m_2 \to m, q$, where $m = m_1 + m_2$ is the total binary mass in units M_0 and $q = m_2/m_1 \le 1$ is the binary mass ratio. Substituting $m_1 = m/(1+q)$, $m_2 = mq/(1+q)$ and the Jacobian $J(m, q) = m/(1+q)^2$ into (4.3) we get

$$\begin{aligned} \frac{d\mathcal{R}(z)}{dmdq} &= \underbrace{\frac{1.6\times 10^6}{\mathrm{Gpc}^3\mathrm{yr}} f_{pbh}^{\frac{53}{57}} \left(\frac{t(z)}{t_0}\right)^{-\frac{34}{37}} \left(\frac{\gamma}{\pi}\right) e^{-\frac{2}{\gamma}} \left(\frac{M_\odot}{M_0}\right)^{-\frac{32}{37}}}_{A} \\ &\quad \times \frac{m^2 q}{(1+q)^2} e^{-\gamma \ln^2(\frac{m}{1+q})} e^{-\gamma \ln^2(\frac{mq}{1+q})} m^{-\frac{32}{37}} \left(\frac{q}{(1+q)^2}\right)^{-\frac{34}{37}} \frac{m}{(1+q)^2} \\ &= A \times m^{(3-\frac{32}{37})} e^{-2\gamma \ln^2 m} \\ &\quad \underbrace{\times\, e^{-\gamma(\ln^2 \frac{1}{1+q} + \ln^2 \frac{q}{1+q})} \left(\frac{q}{(1+q)^2}\right)^{1-\frac{34}{37}-2\gamma \ln m} \frac{1}{(1+q)^2}}_{\Phi(q,m)}. \end{aligned} \tag{4.4}$$

Performing integration over q we obtain the specific rate

$$\frac{d\mathcal{R}(z)}{dm} = A \times m^{(3-\frac{32}{37})} e^{-2\gamma \ln^2 m} \int_0^1 \Phi(q, m) dq\,. \tag{4.5}$$

The specific binary PBH coalescence rate $md\mathcal{R}/dm$ at $z = 0$ for several fiducial model parameters is shown in Fig. 4.1. Note that by adjusting f_{pbh}, at low masses ($M \sim 100 M_\odot$) this rate can be easily made consistent with the observed binary BH merging rate ($\sim 100\,\mathrm{Gpc}^{-3}\mathrm{yr}^{-1}$ at $z \sim 0$) as inferred from LIGO-Virgo-KAGRA observations [45].

The integration of $d\mathcal{R}(z)/dm$ over m yields the total binary PBH merging rate as a function of redshift (Fig. 4.2). Unlike astrophysical binary BH, in the adopted model the binary PBH merging rate monotonically increases with redshift as $\mathcal{R}(z) \sim 1/t(z)$ at all z. We note that the calculated here specific binary IMBH merging rate does not contradict the SMBH merging rate inferred from the recent pulsar timing measurements (see, e.g., [22]).

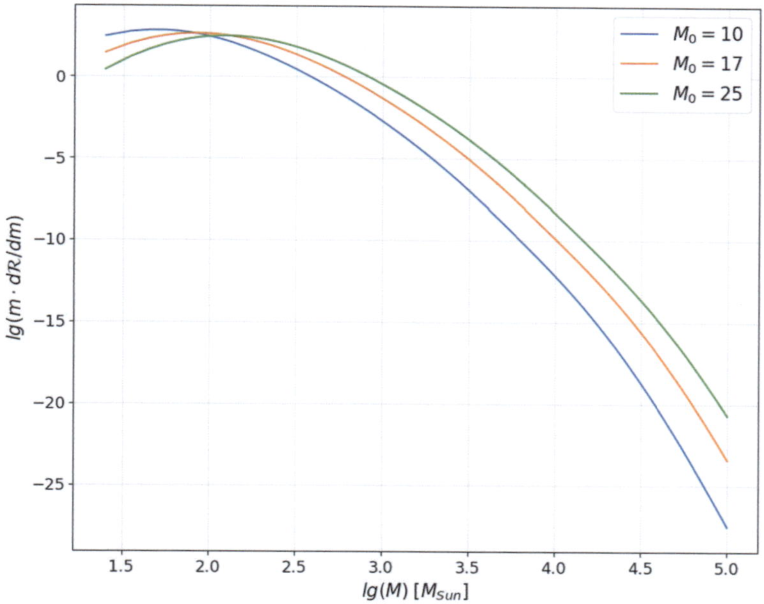

Fig. 4.1 Expected distribution of the primordial binary IMBH merging rate over total mass $md\mathcal{R}/dm$ [$\text{Gpc}^{-3}\text{yr}^{-1}$] at $z = 0$ for $f_{pbh} = 10^{-3}$, $\gamma = 1$ and different $M_0 = 10\,,\,17\,,\,25M_\odot$

4.3 Prospects for Detection of Merging Binary Primordial IMBH by TianQin

The detector sensitivity noise curve imposes a limiting distance (detection horizon) from which a binary with total mass m and mass ratio q[1] can be registered with a given signal-to-noise ratio (SNR). We take the TianQin sensitivity and perform the SNR calculation after [23]. The limiting redshit $z_{lim}(m)$ for different SNR and $q = 0.5$ is shown in Fig. 4.3.

The expected merging rate of binary PBH with taking into account the detector sensitivity is obtained by integrating the specific rate $d\mathcal{R}(z)/dm$ over redshift up to $z_{lim}(m)$:

$$\frac{dN}{dtdm}[\text{yr}^{-1}] = \int_0^{z_{lim}(m)} \frac{d\mathcal{R}(z)}{dm}\frac{1}{(1+z)}\frac{dV(z)}{dz}dz\,, \quad \frac{dN}{dt}(>m) = \int_m^\infty \frac{dN}{dtdm}dm \tag{4.6}$$

The obtained cumulative detection rate $\frac{dN}{dt}(\leq m)$ and $\frac{dN}{dt}(\geq m)$ are shown in Figs. 4.4 and 4.5, respectively. It is seen that with the planned TianQin sensitivity

[1] For binary coalescences, SNR$\propto \mathcal{M}^{5/6}$, where $\mathcal{M} = M\frac{q^{3/5}}{(1+q)^{6/5}}$ is the binary chirp mass, is a function of the binary total mass M and mass ratio q.

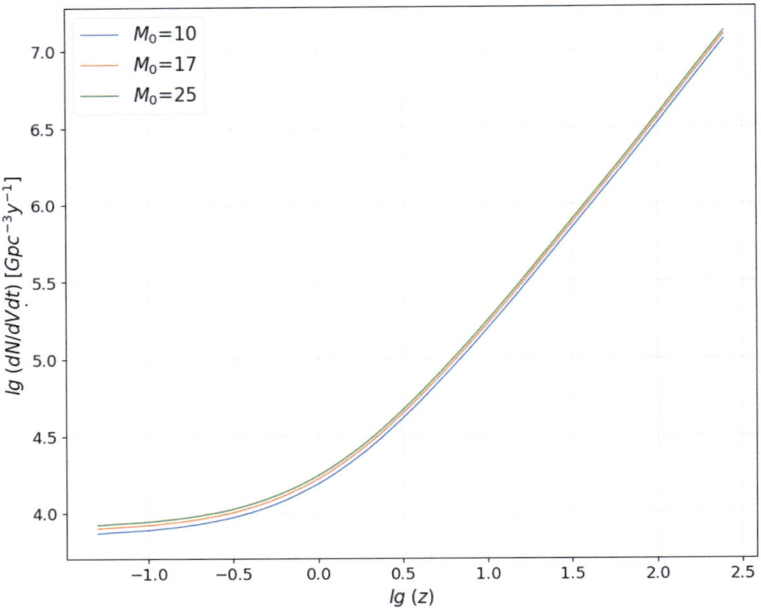

Fig. 4.2 The comoving merging rate of primordial binary BH (4.3) as a function of redhift $\mathcal{R}(z) \sim t(z)^{-34/37}$

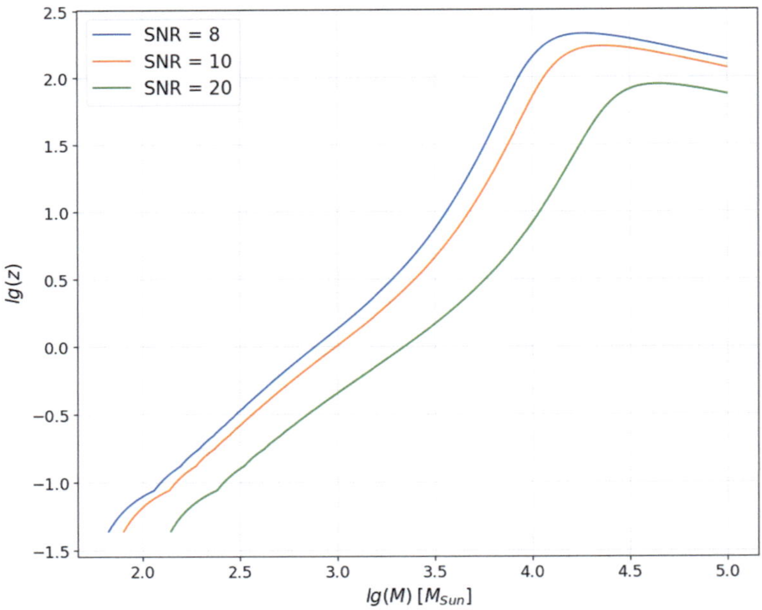

Fig. 4.3 Limiting redshift $z_{lim}(m)$ for detection of coalescing binary BHs with total mass m (in the source's frame) and mass ratio $q = 0.5$ at TianQin threshold SNR=8, 10, 20

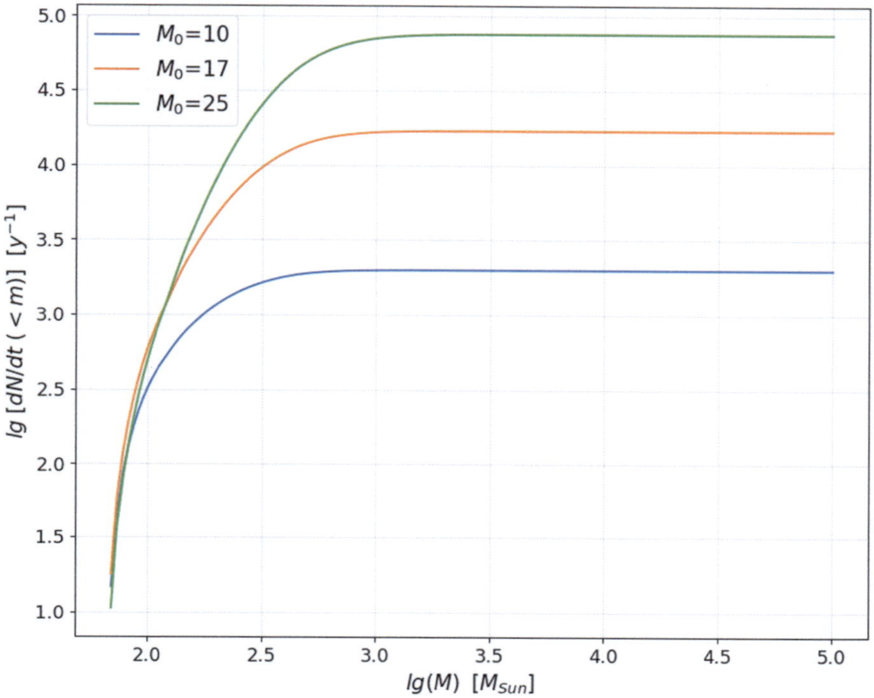

Fig. 4.4 Expected cumulative detection rate $\frac{dN}{dt}(\leq m)$ $[\text{yr}^{-1}]$ of the primordial binary IMBH coalescence rate with total mass m by TianQin interferometer for SNR = 8, $f_{pbh} = 10^{-3}$, $\gamma = 1$

and the adopted assumptions on primordial binary IMBH formation, the detection of a few mergings of primordial binary IMBHs is possible. A unique signature of such mergings would be their zero effective spins and high redshift $z > 20$ where no other but primordial black holes are thought to exist.

4.4 Conclusions

IMBHs with masses from several hundred to tens of thousands of solar masses are very topical in modern astrophysics and cosmology. Their formation and subsequent evolution (accretion mass growth in galactic centers, and coalescence rate of binary IMBHs) remain disputable. Here we have focused on a particular class of IMBHs, primordial IMBHs. These objects can emerge at the radiation-dominated stage in the early Universe during the QCD phase transition from primordial isocurvature fluctuations with large baryonic numbers (HBBs) with the baryon asymmetry parameter $\eta \lesssim 1$, which can be prepared before the end of inflation by a complex scalar

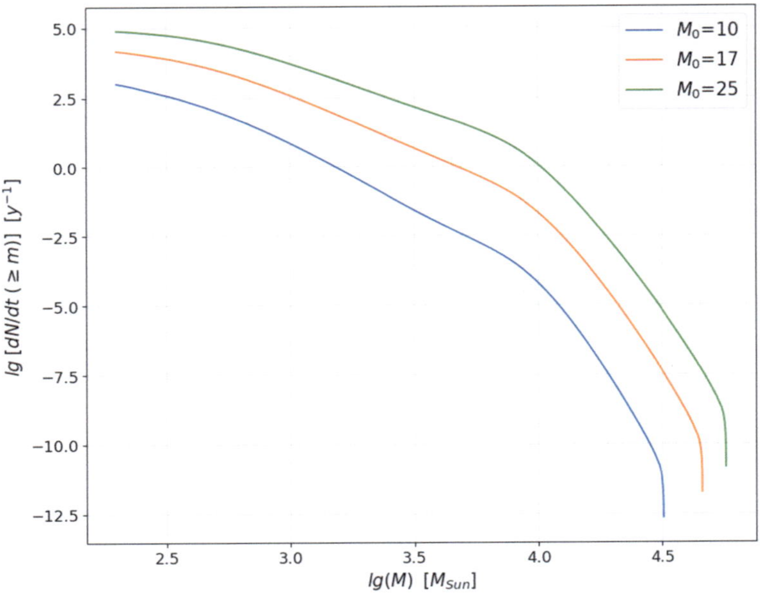

Fig. 4.5 The same as Fig. 4.4 for $\frac{dN}{dt}(\geq m)$

field with baryonic charge coupled to inflation in the modified Affleck-Dine baryogenesis proposed in [17]. This mechanism predicts the existence of a log-normal distribution of the baryonic mass of HBB they acquire after the QCD phase transition at $T_{QCD} \sim 100$ MeV. The expected non-zero chemical potential inside HBBs $\mu \sim 0.3$ GeV decreases T_{QCD}. Large density fluctuations emerging from HBBs after the QCD phase transition can give rise to primordial black holes with a characteristic mass of the order of the cosmological horizon at that time of about $10 - 20 M_\odot$ and a log-normal mass distribution extending up to $\sim 10^4 M_\odot$ [8].

Using a particular model of primordial binary BH formation [41, 42], we calculated the expected detection rate of such binary PBHs by the TianQin interferometer (Figs. 4.4, 4.5). It is shown that for PBH formation parameters consistent with the detection rate and distribution of binary BH mergings by the LIGO-Virgo-KAGRA interferometers ($M_0 \sim 10 - 20 M_\odot, \gamma \sim 1$) and a modest PBH fraction in dark matter density $f_{pbh} = 10^{-3}$ [4, 20, 30, 34], the expected number of detections of merging IMBHs with total mass $M \sim 10^3 - 10^4 M_\odot$ can range from a few to hundred events in several years of observations depending on parameters. We caution that this estimate is based on several model assumptions, including the shape of the PBH initial mass spectrum [28], the PBH fraction, and possible modification of the binary PBH merging rate due to additional binary formation in the large-scale structure environment in the later matter-dominated stage [3, 14]. We conclude that space laser gravitational-wave interferometers sensitive at mHz like TianQin will be able to prove the existence of primordial IMBHs having the unique feature of being only binary IMBHs mergings at high redshifts $z > 20$ in the dark age epoch. Even at lower redshifts in the stellar epoch, the mergings of primordial IMBHs can not be associated with gas-rich regions or galaxies and thus are not expected to be associated with strong electromagnetic signals.

Acknowledgements We thank A. D. Dolgov for discussions. The authors acknowledge the support from the Russian Science Foundation grant 23-42-00055. KP also thanks the TianQin Research Center for Gravitational Physics in Zhuhai for hospitality.

References

1. B.P. Abbott, R. Abbott, T.D. Abbott, M.R. Abernathy, F. Acernese, K. Ackley, C. Adams, T. Adams, P. Addesso, R.X. Adhikari et al., Observation of gravitational waves from a binary black hole merger. Phys. Rev. Lett. **116**(6), 061102 (2016)
2. R. Abbott, T.D. Abbott, Acernese, et al., Gwtc-3: compact binary coalescences observed by ligo and virgo during the second part of the third observing run. Phys. Rev. X **13**, 041039 (2023)
3. M. Aljaf, I. Cholis, Simulating Binary Primordial Black Hole Mergers in Dark Matter Halos. (2024), arXiv:2408.06515
4. M. Andrés-Carcasona, A.J. Iovino, V. Vaskonen, H. Veermäe, M. Martínez, O. Pujolàs, L.M. Mir, Constraints on primordial black holes from LIGO-Virgo-KAGRA O3 events. (2024), arXiv:2405.05732
5. I.Y. Aref'eva, A. Ermakov, K. Rannu, P. Slepov, Holographic model for light quarks in anisotropic hot dense QGP with external magnetic field. Eur. Phys. J. C **83**(1), 79 (2023)
6. A. Askar, V.F. Baldassare, M. Mezcua, Intermediate-Mass Black Holes in Star Clusters and Dwarf Galaxies (2023), arXiv:2311.12118
7. S. Bird, I. Cholis, J.B. Muñoz, Y. Ali-Haïmoud, M. Kamionkowski, E.D. Kovetz, A. Raccanelli, A.G. Riess, Did LIGO detect dark matter? Phys. Rev. Lett. **116**(20), 201301 (2016)
8. S. Blinnikov, A. Dolgov, N.K. Porayko, K. Postnov, Solving puzzles of GW150914 by primordial black holes. JCAP **2016**(11), 036 (2016)
9. B. Carr, J. Silk, Primordial black holes as generators of cosmic structures. MNRAS **478**(3), 3756–3775 (2018)
10. B.J. Carr, The primordial black hole mass spectrum. ApJ **201**, 1–19 (1975)
11. B.J. Carr, S. Clesse, J. García-Bellido, M. Hawkins, F. Kühnel, Observational evidence for primordial black holes: a positivist perspective. Phys. Rep. **1054**, 1–68 (2024)
12. B.J. Carr, S.W. Hawking, Black holes in the early Universe. MNRAS **168**, 399–416 (1974)
13. P.E. Colazo, F. Stasyszyn, N. Padilla, Structure formation with primordial black holes to alleviate early star formation tension revealed by JWST. A&A **685**, L8 (2024)
14. M.S. Delos, A. Rantala, S. Young, F. Schmidt, Structure formation with primordial black holes: collisional dynamics, binaries, and gravitational waves (2024), arXiv:2410.01876
15. A. Dolgov, K. Postnov, Globular cluster seeding by primordial black hole population. JCAP **2017**(4), 036 (2017)
16. A. Dolgov, K. Postnov, Why the mean mass of primordial black hole distribution is close to $10M_{solar}$. JCAP **2020**(7), 063 (2020)
17. A. Dolgov, J. Silk, Baryon isocurvature fluctuations at small scales and baryonic dark matter. Phys. Rev. D **47**(10), 4244–4255 (1993)
18. A.D. Dolgov, Massive and supermassive black holes in the contemporary and early Universe and problems in cosmology and astrophysics. Phys. Uspekhi **61**(2), 115 (2018)
19. A.D. Dolgov, M. Kawasaki, N. Kevlishvili, Inhomogeneous baryogenesis, cosmic antimatter, and dark matter. Nucl. Phys. B **807**(1–2), 229–250 (2009)

20. A.D. Dolgov, A.G. Kuranov, N.A. Mitichkin, S. Porey, K.A. Postnov, O.S. Sazhina, I.V. Simkin, On mass distribution of coalescing black holes. JCAP **2020**(12), 017 (2020)
21. T. Eftekhari, A. Tchekhovskoy, K.D. Alexander, E. Berger, R. Chornock, T. Laskar, R. Margutti, Y. Yao, Y. Cendes, S. Gomez, A. Hajela, D.R. Pasham, Late-time X-ray Observations of the Jetted Tidal Disruption Event AT2022cmc: The Relativistic Jet Shuts Off (2024), arXiv:2404.10036
22. EPTA Collaboration, InPTA Collaboration, J. Antoniadis, P. Arumugam, S. Arumugam et.al., The second data release from the European Pulsar Timing Array. IV. Implications for massive black holes, dark matter, and the early Universe. A&A **685**, A94 (2024)
23. W.-F. Feng, H.-T. Wang, X.-C. Hu, Y.-M. Hu, Y. Wang, Preliminary study on parameter estimation accuracy of supermassive black hole binary inspirals for TianQin. Phys. Rev. D **99**(12), 123002 (2019)
24. G. Fragione, A. Loeb, Constraining the cosmic merger history of intermediate-mass black holes with gravitational wave detectors. ApJ **944**(1), 81 (2023)
25. M.S. Fujii, L. Wang, A. Tanikawa, Y. Hirai, T.R. Saitoh, Simulations predict intermediate-mass black hole formation in globular clusters. Science (2024)
26. E. González Prieto, N.C. Weatherford, G. Fragione, K. Kremer, F.A. Rasio, IMBH Progenitors from Stellar Collisions in Dense Star Clusters (2024), arXiv:2404.11646
27. M. Häberle, N. Neumayer, A. Seth, A. Bellini, M. Libralato, H. Baumgardt, M. Whitaker, A. Dumont, M. Alfaro Cuello, J. Anderson, C. Clontz, N. Kacharov, S. Kamann, A. Feldmeier-Krause, A. Milone, M.S. Nitschai, R. Pechetti, G. van de Ven, Fast-moving stars around an intermediate-mass black hole in Omega Centauri (2024), arXiv:2405.06015
28. A. Hall, A.D. Gow, C.T. Byrnes, Bayesian analysis of LIGO-Virgo mergers: primordial versus astrophysical black hole populations. Phys. Rev. D **102**(12), 123524 (2020)
29. Y. Huang, Q. Li, J. Liu, X. Dong, H. Zhang, Y. Lu, C. Du, A high-velocity star recently ejected by an intermediate-mass black hole in M15 (2024), arXiv:2406.00923
30. G. Hütsi, M. Raidal, V. Vaskonen, H. Veermäe, Two populations of LIGO-Virgo black holes. JCAP **2021**(3), 068 (2021)
31. J. Kormendy, L.C. Ho, Coevolution (or not) of supermassive black holes and host galaxies. ARAA **51**(1), 511–653 (2013)
32. O.E. Kovács, Á. Bogdán, P. Natarajan, N. Werner, M. Azadi, M. Volonteri, G.R. Tremblay, U. Chadayammuri, W.R. Forman, C. Jones, R.P. Kraft, A candidate supermassive black hole in a gravitationally lensed galaxy at Z $\approx$ 10. ApJL **965**(2), L21 (2024)
33. E.-K. Li, S. Liu, A. Torres-Orjuela et al., Gravitational Wave Astronomy With TianQin (2024), arXiv:2409.19665
34. B. Liu, V. Bromm, Impact of primordial black holes on the formation of the first stars and galaxies (2023), arXiv:2312.04085
35. L. Liu, Z.-Q. You, Y. Wu, Z.-C. Chen, Constraining the merger history of primordial-black-hole binaries from GWTC-3. Phys. Rev. D **107**(6), 063035 (2023)
36. J. Luo et al., TianQin: a space-borne gravitational wave detector. Class. Quant. Grav. **33**(3), 035010 (2016)
37. J. Mei et al., The TianQin project: current progress on science and technology. PTEP **2021**(5), 05A107 (2021)
38. M. Mezcua, Observational evidence for intermediate-mass black holes. Int. J. Mod. Phys. D **26**(11), 1730021 (2017)
39. I.F. Mirabel, L.F. Rodríguez, Black holes at cosmic dawn in the redshifted 21cm signal of HI. New Astron. Rev. **94**, 101642 (2022)
40. R. Pascale, C. Nipoti, F. Calura, A. Della Croce, The central black hole in the dwarf spheroidal galaxy Leo I: Not supermassive, at most an intermediate-mass candidate. A&A **684**, L19 (2024)
41. M. Raidal, C. Spethmann, V. Vaskonen, H. Veermäe, Formation and evolution of primordial black hole binaries in the early universe. JCAP **2019**(2), 018 (2019)
42. M. Raidal, V. Vaskonen, H. Veermäe, Formation of primordial black hole binaries and their merger rates (2024), arXiv:2404.08416

43. M. Sasaki, T. Suyama, T. Tanaka, S. Yokoyama, Primordial black hole scenario for the gravitational-wave event GW150914. Phys. Rev. Lett. **117**(6), 061101 (2016)
44. R. Schneider, A. Ferrara, P. Natarajan, K. Omukai, First stars, very massive black holes, and metals. ApJ **571**(1), 30–39 (2002)
45. The LIGO Scientific Collaboration, Population of merging compact binaries inferred using gravitational waves through gwtc-3. Phys. Rev. X **13**, 011048 (2023)
46. S. Young, Computing the abundance of primordial black holes (2024), arXiv:2405.13259
47. Y.B. Zel'dovich, I.D. Novikov, The hypothesis of cores retarded during expansion and the hot cosmological model. Sov. Astron. **10**, 602 (1967)

Chapter 5
Gravitational Wave and Radio Observations of Galactic Inspiraling Double Neutron Stars

Wen-Fan Feng, Tan Liu, Jie-Wen Chen, Yan Wang, Soumya D. Mohanty, and Yong Shao

Abstract We integrate the key results of our previous studies [8, 9, 11], providing a unified view on multimessenger and multiband (dual-line) observations of inspiraling double neutron stars (DNSs) in our Galaxy. Future space-based gravitational wave (GW) detectors, such as LISA, TianQin, and Taiji, are poised to bridge the detection

W.-F. Feng (✉) · T. Liu · J.-W. Chen · Y. Wang
National Gravitation Laboratory, MOE Key Laboratory of Fundamental Physical Quantities Measurements, Department of Astronomy and School of Physics, Huazhong University of Science and Technology, Wuhan, China
e-mail: fengwf@pku.edu.cn

T. Liu
e-mail: lewton@mail.ustc.edu.cn

Y. Wang
e-mail: ywang12@hust.edu.cn

W.-F. Feng
Kavli Institute for Astronomy and Astrophysics, Peking University, Beijing, China

T. Liu
School of Fundamental Physics and Mathematical Sciences, Hangzhou Institute for Advanced Study, UCAS, Hangzhou, China

University of Chinese Academy of Sciences, Beijing, China

J.-W. Chen
National Time Service Center, Chinese Academy of Sciences, Xi'an, China

Key Laboratory of Time and Frequency Primary Standards, Chinese Academy of Sciences, Xi'an, China

S. D. Mohanty
Department of Physics and Astronomy, University of Texas Rio Grande Valley, Brownsville, USA

Department of Physics, IIT Hyderabad, Kandai, Telangana, India

Y. Shao
School of Astronomy and Space Science, Nanjing University, Nanjing, China

Key Laboratory of Modern Astronomy and Astrophysics, Nanjing University, Ministry of Education, Nanjing, China

J. Luo et al. (eds.), *Proceedings of the 7th International Workshop on the TianQin Science Mission; 25–26 April 2024, Hong Kong SAR, China*, Springer Proceedings in Physics 428, https://doi.org/10.1007/978-981-96-7803-7_5

gap of tight DNSs with orbital periods of approximately 10 min, which remain challenging to detect with radio telescopes. Our investigation will first explore GW and radio follow-up detection capabilities for Galactic inspiraling DNSs. Furthermore, next-generation ground-based GW observatories, such as Cosmic Explorer and Einstein Telescope, which are projected to be operational in the mid-2030s concurrently with LISA, TianQin, and Taiji, are expected to detect high-frequency GWs emitted by spinning NSs. This development will advance the study of dual-line GWs from DNSs. We then focus on GW waveform modeling for the spinning NS in a tight DNS and consider its potential role in inferring binary geometry and NS physics parameters by dual-line GW detection.

5.1 Introduction

Binary pulsars, known for emitting both radio and gravitational waves (GWs), serve as invaluable laboratories for testing gravitational theories, a utility recognized since the discovery of PSR B1913+16 [16, 17]. Current radio surveys, constrained by various technical limitations, can only detect binary systems with orbital periods exceeding approximately one hour. However, forthcoming space-based GW detectors, such as LISA [2], TianQin (TQ) [21, 25], and Taiji [14], will be able to identify systems with much shorter orbital periods. This advancement raises interest in the potential for radio follow-up observations of short-period double neutron star (DNS) systems detected by GWs.

On the other hand, DNSs can serve as intriguing sources of GWs because of the distinct emissions (dual lines) resulting from both the potential asymmetry of the individual stars, which produces high-frequency waves, and the orbital motion of the binary system, which produces low-frequency waves. The next generation of ground-based GW detectors, such as Cosmic Explorer [30] and Einstein Telescope [28], operating concurrently with LISA, TQ, and Taiji, are expected to make dual-line GW detections a reality. Such multiband detections will improve our understanding of the orbital geometries of binaries and the physical properties of NSs.

Previous works have focused on the detectability of DNSs in our Galaxy and nearby galaxies by LISA alone [3, 20, 29]. To address the challenges of detecting short-period DNSs by radio observations, a multimessenger strategy is proposed to detect Galactic pulsar-NS binaries with orbital periods less than 10 min [19]. In this strategy, the sky localization and orbital parameters accurately determined by LISA can facilitate pulsar searches by the Square Kilometer Array (SKA) [6]. Regarding the prospects for dual-line GW detection, the high-frequency GWs emitted by spinning NS in a binary system are still treated as the waveforms of an isolated spinning NS modulated by Doppler frequency shift [1]. However, for short-period systems detectable by LISA, TQ, or Taiji, interactions between the NS and its companion, such as spin-orbit coupling [27], must be taken into account.

This paper aims to integrate insights from three of our previous studies [8, 9, 11], providing a unified view of multimessenger and multiband observations of inspiraling

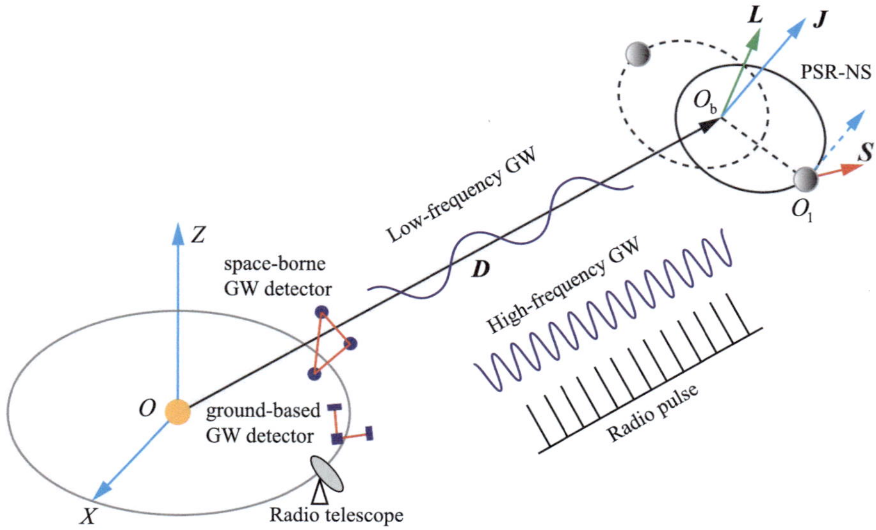

Fig. 5.1 Illustration of multimessenger and multiband observations of an inspiraling DNS consisting of a pulsar and an NS with no visible radio emission (modified from Fig. 1 in [8])

DNSs in the Galactic disk. We attempt to offer new insights by integrating the results and discussing them in a broader context. Using TQ and its combination with LISA, we investigate the detectability of these DNSs and estimate the parameter estimation accuracies by considering the antenna responses of TQ and LISA. We also explore the follow-up search for potential radio pulsars in the DNSs based on the characteristics of pulsars and the specifications of radio telescopes, including the Parkes radio telescope [13], the Five-hundred-meter Aperture Spherical radio Telescope (FAST) [26], the SKA Phase 1 (SKA1) [5], and the SKA [32]. In addition, we estimate the timing precision of the pulsars detectable in DNSs by these radio telescopes, a critical factor for further investigations of tight DNSs as laboratories for astrophysics. The advancement of next-generation GW detectors has made it possible to realize dual-line GW detections. Therefore, we calculate the GWs emitted by spinning NSs in the TQ/LISA binaries, incorporating spin-orbit coupling effects. Furthermore, we explore the potential applications of these waveforms in the detection of dual-line GWs (Fig. 5.1).

5.2 Multimessenger Detection of Galactic Inspiraling DNSs

We first synthesize the Galactic DNS population following the method in [3]. The composition of the population consists of three main parts: First, based on the expected merger time of the fast-merging channel and the predicted merger rate from DNS GW detections made by ground-based detectors, we obtain the simulated number of the DNSs. Then, by setting the merger time and orbital parameters close to the merger, we calculate the orbital parameters of the DNSs according to the evolution equations. Finally, we assign these sources to the positions given by the Galactic disk model to generate the DNS population.

Here we use the waveforms of inspiraling binaries in eccentric orbits, combined with the antenna response of the detectors. The Fisher information matrix is used to evaluate the estimation accuracies for the unknown parameters that characterize the systems, which shows that TQ, when combined with LISA, can achieve a sky localization accuracy of $< 100\ \mathrm{deg}^2$, and relative accuracies for eccentricity, orbital period, and period derivative of < 0.3, $< 10^{-6}$, and < 0.02, respectively. Note that the values mentioned above refer to the upper boundary of detection. The localizations of DNSs can be as precise as $10^{-3}\ \mathrm{deg}^2$ for the optimal signal-to-noise ratio (SNR) $\sim 10^3$ (see Fig. 3 in Ref. [9]). This contrasts with the localizations for stellar-mass binary black holes ($[10^{-2}, 10^2]\ \mathrm{deg}^2$) and double white dwarfs ($[10^{-6}, 10^2]\ \mathrm{deg}^2$) by TQ [15, 22]. Through 100 Monte Carlo simulations, TQ, LISA, and their combination are expected to detect 217, 368, and 429 DNSs on average, respectively, with a SNR of the GW signal greater than 7.

Next, based on the strategy of GW-assisted radio observations, we model the radio pulse radiation for the simulated DNS population. Adopting the magnetic dipole radiation model, pulses are only observable when the angle between the observer's line of sight and the magnetic axis is smaller than the radius of the radiation cone. This constraint allows for the calculation of the pulse width [24]. In our simulation, we assume that each DNS is composed of a millisecond pulsar (MSP) and a normal pulsar (NP). Based on the empirical distribution formula for pulsars, each NS is assigned appropriate parameters, such as spin period, luminosity, and the radius of the radiation cone [7, 18, 23].

By combining the technical specifications and observation parameters of different radio telescopes (Parkes, FAST, SKA1-Mid, and SKA-Mid), we calculate the SNRs of the integrated pulse profiles for the pulsar population. This calculation takes into account the declination ranges of different telescopes. By conducting 100 Monte Carlo simulations, we find that for TQ+LISA DNSs, follow-up pulsar searches with Parkes, FAST, SKA1, and SKA can detect an average of 8, 10, 43, and 87 DNSs, respectively. The average distances of these binaries range from 1 to 7 kpc, depending on the radio telescopes. The timing accuracy of these pulsars can be as low as 70 ns by calculating the dominant noise sources, namely radiometer and jitter noises.

5.3 Dual-Line GW Modeling for Inspiraling DNSs and Its Implications

In this study, we model the GWs emitted by spinning NSs within circular binaries, including the effects of spin-orbit coupling. Our approach extends the model from triaxial aligned NSs to general triaxial nonaligned NSs. Following methodologies established in previous research, such as those used by LIGO [1], we derive the GWs produced by spinning NSs in a binary system. This is achieved by incorporating the Doppler frequency modulation into the phase of the waveforms for triaxial non-aligned NSs undergoing spin precession. In [11], we neglect additional influences such as the magnetic dipole field, tidal interactions, or radiation reaction effects. In contrast to the isolated NS scenario, where the spin angular momentum remains constant when electromagnetic and gravitational radiation-reaction torques are neglected, in a binary setting, the spin angular momentum of the NS experiences precession due to spin-orbit coupling [4].

We analytically resolve the spin precession dynamics for an NS, using perturbation methods to determine the evolution of the spin angular frequency in [11]. The resulting waveforms are then calculated using the quadrupole moment approach. Further refinement is achieved by decomposing these waveforms into a set of components in the regime of small parameters, facilitating continuous GW detectability analysis. Utilizing these accessible waveform components, we inspect the influence of spin-orbit coupling effects on the accuracy of parameter estimation for spinning NSs.

Our results for an NS in a tight DNS system are twofold: (i) The waveforms of an NS undergoing spin precession begin to diverge significantly from those of an isolated star within a few days ($\approx 10^5$ seconds), as evidenced by the fitting factor between the two waveforms dropping below 0.97. (ii) Spin-orbit coupling can significantly improve the accuracy of parameter estimation. This is particularly evident in the determination of the binary inclination angle and the spin precession cone opening angle, with potential improvements reaching up to three orders of magnitude.

These improved GW models contain more information about binary geometry and NS physics than isolated cases. Without relying on electromagnetic observations, we have developed an approach to infer the spin-orbit misalignment angle by considering tight dual-line DNSs influenced by spin-orbit coupling. Using a combined four-year LISA and the Cosmic Explorer observing campaign, we find that it is possible to determine the misalignment angle and the moment of inertia of the NS with sub-percentage and 5% accuracy, respectively.

Subsequently, we further extend the analytical waveforms for NSs to include the effects of orbital eccentricity [12]. Our results show that for DNSs formed by isolated binary evolution, the influence of eccentricity on the continuous GWs from their NSs can be disregarded. In contrast, for GWs from NSs in DNSs generated through dynamical processes, it is important to account for eccentricity, as orbits with high eccentricity can lead to a fitting factor of <0.97 within 0.5–2 years of observation at GW frequencies of 100 Hz and 200 Hz.

The number of detectable double-line DNSs can be estimated by modeling the spinning NSs within these systems. Using COMPAS population synthesis simulations of LISA DNSs [31], we conduct a Monte Carlo simulation for the structural parameters of the spinning NSs. Then, by employing Cosmic Explorer for NS detection (with SNR threshold also set to 7), we determine the number of double-line GW detections based on both LISA and Cosmic Explorer. Preliminary results indicate that if LISA can detect 35 pairs of DNSs over a 4-year observation period, the number of double-line detections using Cosmic Explorer would be 6(22) for log-uniform (uniform) sampling of the NSs' characteristic parameters [10]. The number is expected to be even larger for those DNSs formed through the fast-merging channel studied in this paper.

5.4 Conclusions

This work evaluates the detection capabilities of Galactic DNSs with TQ and its combination with LISA, and investigates the prospects for radio follow-up of pulsar components in these GW-detected DNSs. These simulations provide observational strategies for searches with radio telescopes. Improved dual-line GW modeling of LISA/TQ binaries contributes to a deeper understanding of binary orbits and the internal structure of NSs. Especially, we found that the spin-orbit coupling can potentially improve the accuracy of the parameter estimation for tight DNS systems.

Acknowledgements Y.W. gratefully acknowledges support from the National Key Research and Development Program of China, the National Natural Science Foundation of China (NSFC) under Grants No. 11973024, Major Science and Technology Program of Xinjiang Uygur Autonomous Region (No. 2022A03013-4), and Guangdong Major Project of Basic and Applied Basic Research (Grant No. 2019B030302001). T. L. is supported by NSFC Grant No. 12003008. J.-W. C. acknowledges the support from China Postdoctoral Science Foundation under Grant No. 2021M691146. S.D.M is supported by U.S. National Science Foundation (NSF) grant PHY-2207935.

References

1. R. Abbott, T.D. Abbott et al., All-sky search in early o3 ligo data for continuous gravitational-wave signals from unknown neutron stars in binary systems. Phys. Rev. D **103**, 064017 (2021)
2. P. Amaro-Seoane, H. Audley, S. Babak, J. Baker, E. Barausse, P. Bender, E. Berti, P. Binetruy, M. Born, D. Bortoluzzi et al., Laser interferometer space antenna (2017). arXiv:1702.00786
3. J.J. Andrews, K. Breivik, C. Pankow, D.J. D'Orazio, M. Safarzadeh, Lisa and the existence of a fast-merging double neutron star formation channel. Astrophys. J. Lett. **892**(1), L9 (2020)
4. T.A. Apostolatos, C. Cutler, G.J. Sussman, K.S. Thorne, Spin-induced orbital precession and its modulation of the gravitational waveforms from merging binaries. Phys. Rev. D **49**, 6274–6297 (1994)
5. P. Dewdney, W. Turner, R. Braun, J. Santander-Vela, M. Waterson, G.-H. Tan, *SKA1 System Baseline Design* (Square Kilometre Array Organisation, Jodrell Bank, UK, 2016)
6. P.E. Dewdney, P.J. Hall, R.T. Schilizzi, T. Lazio, The square kilometre array. IEEE Proc. **97**(8), 1482–1496 (2009)
7. C.-A. Faucher-Giguère, V.M. Kaspi, Birth and evolution of isolated radio pulsars. Astrophys. J. **643**(1), 332–355 (2006)
8. W.-F. Feng, J.-W. Chen, T. Liu, Y. Wang, S.D. Mohanty, Inferring binary parameters with dual-line gravitational wave detection from tight inspiraling double neutron stars. Phys. Rev. D **109**, 043033 (2024)

9. W.-F. Feng, J.-W. Chen, Y. Wang, S.D. Mohanty, Y. Shao, Multimessenger observations of double neutron stars in the galactic disk with gravitational and radio waves. Phys. Rev. D **107**, 103035 (2023)
10. W.-F. Feng, L. Shao, Galactic double neutron stars as dual-line gravitational-wave sources: Prospects with LISA and Cosmic Explorer. Phys. Rev. D **112**, 043023 (2025)
11. W.-F. Feng, T. Liu, J.-W. Chen, Y. Wang, S.D. Mohanty, Effects of spin-orbit coupling on gravitational waveforms from a triaxial nonaligned neutron star in a binary system. Phys. Rev. D **108**, 063035 (2023)
12. W.-F. Feng, T. Liu, Y. Wang, L. Shao, Effects of orbital eccentricity on continuous gravitational waveforms from triaxially-deformed precessing neutron stars in tight binaries. Phys. Rev. D. **111**, 023053 (2025)
13. G. Hobbs, R.N. Manchester et al., An ultra-wide bandwidth (704 to 4 032 mhz) receiver for the parkes radio telescope. Publ. Astron. Soc. Austral. **37**, e012 (2020)
14. W.-R. Hu, Y.-L. Wu, The taiji program in space for gravitational wave physics and the nature of gravity. Natl. Sci. Rev. **4**(5), 685–686 (2017)
15. S.-J. Huang, Y.-M. Hu, V. Korol, P.-C. Li, Z.-C. Liang, Y. Lu, H.-T. Wang, S. Yu, J. Mei, Science with the tianqin observatory: preliminary results on galactic double white dwarf binaries. Phys. Rev. D **102**, 063021 (2020)
16. R.A. Hulse, J.H. Taylor, Discovery of a pulsar in a binary system. Astrophys. J. Lett. **195**, L51–L53 (1975)
17. M. Kramer, I.H. Stairs et al., Strong-field gravity tests with the double pulsar. Phys. Rev. X **11**(4), 041050 (2021)
18. M. Kramer, K.M., Xilouris, D.R. Lorimer, O. Doroshenko, A. Jessner, R. Wielebinski, A. Wolszczan, F. Camilo, The characteristics of millisecond pulsar emission. I. Spectra, pulse shapes, and the beaming fraction. Astrophys. J. **501**(1), 270–285 (1998)
19. K. Kyutoku, Y. Nishino, N. Seto, How to detect the shortest period binary pulsars in the era of lisa. Mon. Not. R. Astron. Soc. **483**(2), 2615–2620 (2019)
20. M. Lau, I. Mandel, A. Vigna-Gómez, C.J. Neijssel, S. Stevenson, A. Sesana, Detecting double neutron stars with lisa. Mon. Not. R. Astron. Soc. **492**(3), 3061–3072 (2020)
21. E.-K. Li, S. Liu, A. Torres-Orjuela, X. Chen, K. Inayoshi, L. Wang, Y.-M. Hu, P. Amaro-Seoane, A. Askar, C. Bambi, P.R. Capelo, H.-Y. Chen, A.J.K. Chua, E. Condés-Breña, L. Dai, D. Das, A. Derdzinski, H.-M. Fan, M. Fujii, J. Gao, M. Garg, H. Ge, M. Giersz, S.-J. Huang, A. Hypki, Z.-C. Liang, B. Liu, D. Liu, M. Liu, Y. Liu, L. Mayer, N.R. Napolitano, P. Peng, Y. Shao, S. Shashank, R. Shen, H. Tagawa, A. Tanikawa, M. Toscani, V. Vázquez-Aceves, H.-T. Wang, H. Wang, S.-X. Yi, J.-D. Zhang, X.-T. Zhang, L. Zhu, L. Zwick, S. Huang, J. Mei, Y. Wang, Y. Xie, J. Zhang, J. Luo, Gravitational Wave Astronomy With TianQin. Rept. Prog. Phys. **88**, 056901 (2025)
22. S. Liu, Y.-M. Hu, J.-D. Zhang, J. Mei, Science with the tianqin observatory: preliminary results on stellar-mass binary black holes. Phys. Rev. D **101**, 103027 (2020)
23. D.R. Lorimer, A.J. Faulkner, A.G. Lyne, R.N. Manchester, M. Kramer, M.A. McLaughlin, G. Hobbs, A. Possenti, I.H. Stairs, F. Camilo, M. Burgay, N. D'Amico, A. Corongiu, F. Crawford, The Parkes Multibeam Pulsar Survey - VI. Discovery and timing of 142 pulsars and a Galactic population analysis. Mon. Notices Royal Astron. Soc. **372**(2), 777–800 (2006)
24. D.R. Lorimer, M. Kramer, *Handbook of Pulsar Astronomy*, vol. 4 (Cambridge University Press, 2004)
25. J. Luo et al., Tianqin: a space-borne gravitational wave detector. Class. Quant. Grav. **33**(3), 035010 (2016)
26. R. Nan, D. Li, C. Jin, Q. Wang, L. Zhu, W. Zhu, H. Zhang, Y. Yue, L. Qian, The five-hundred aperture spherical radio telescope (fast) project. Int. J. Mod. Phys. D **20**(6), 989–1024 (2011)
27. E. Poisson, C.M. Will, Gravity: Newtonian, Post-Newtonian, Relativistic (Cambridge University Press, 2014)
28. M. Punturo, M. Abernathy et al., The einstein telescope: a third-generation gravitational wave observatory. Class. Quantum Gravity **27**(19), 194002 (2010)

29. N. Seto, Search for neutron star binaries in the local group galaxies using lisa. Mon. Not. R. Astron. Soc. **489**(4), 4513–4519 (2019)
30. V. Srivastava, D. Davis, K. Kuns, P. Landry, S. Ballmer, M. Evans, E.D. Hall, J. Read, B.S. Sathyaprakash, Science-driven tunable design of cosmic explorer detectors. Astrophys. J. **931**(1), 22 (2022)
31. T. Wagg, F.S. Broekgaarden, S.E. de Mink, N. Frankel, L. van Son, S. Justham, Gravitational wave sources in our galactic backyard: predictions for bhbh, bhns, and nsns binaries detectable with lisa. Astrophys. J. **937**(2), 118 (2022)
32. A. Weltman, P. Bull et al., Fundamental physics with the square kilometre array. Publ. Astron. Soc. Austral. **37**, e002 (2020)

Chapter 6
Testing the Kerr Nature with Binary Black Hole Inspirals

Swarnim Shashank, Cosimo Bambi, and Rittick Roy

Abstract The theory of general relativity (GR) is the standard framework for the description of gravitation and the geometric structure of spacetime. With the recent advancement of observational instruments, it has become possible to probe the strong field regime to test GR. We present the constraints obtained from the binary black hole inspiral data of the LIGO-Virgo-Kagra (LVK) gravitational wave (GW) observations on the deformations of some popular parametrized non-Kerr metrics.

6.1 Introduction

Shortly after formulating general relativity (GR), Einstein derived the emission of gravitational radiation [13], now popularly termed gravitational waves (GWs). After 100 years, in 2015, the first detection of GWs from the coalescence of two black holes (BHs) was made by the LIGO labs [1], and since then, around 100 events have been observed [2, 4, 7]. Using the binary BH data observed by the LIGO-Virgo-Kagra (LVK) collaboration, tests of GR can be performed [3, 5, 6]. One such test is the test of the Kerr hypothesis. The Kerr hypothesis states that astrophysical black holes are Kerr in nature [16]. Testing the Kerr nature of a black hole is usually performed by assuming a parametrized spacetime where, in addition to the mass and spin, there are extra non-Kerr parameters that show a deviation from the Kerr geometry (see, e.g., [9, 14, 15, 17, 23, 26, 29, 30]). Recent efforts have focused on testing the Kerr hypothesis using electromagnetic spectrum data from X-ray and radio observations [8, 10, 25, 33] and GW observations [11, 26, 29, 30].

S. Shashank (✉) · C. Bambi
Center for Astronomy and Astrophysics, Center for Field Theory and Particle Physics and Department of Physics, Fudan University, Shanghai, People's Republic of China
e-mail: swarnim@fudan.edu.cn

C. Bambi
School of Humanities and Natural Sciences, New Uzbekistan University, Tashkent, Uzbekistan

R. Roy
Anton Pannekoek Institute for Astronomy, University of Amsterdam, Amsterdam, The Netherlands

J. Luo et al. (eds.), *Proceedings of the 7th International Workshop on the TianQin Science Mission; 25–26 April 2024, Hong Kong SAR, China*, Springer Proceedings in Physics 428, https://doi.org/10.1007/978-981-96-7803-7_6

This contribution is based on [26, 30], where GW data from the events observed by the LVK are used to constrain the non-Kerr parameters of some popular metrics. In the rest of the article, we dedicate each section to one metric. We very briefly explain the metric and its deformation parameters and then present the constraints obtained from the LVK data. We use the publicly available posterior samples that were obtained using Bayesian analysis [19, 20].

6.2 KRZ Metric

The Konoplya, Rezzolla & Zhidenko (KRZ) metric [17] is one of the popular parametrized axisymmetric metrics studied in the literature. In our methods, we study the inspirals of the GW sources; hence, we only consider equatorial circular orbits ($\theta = \pi/2$ and $\dot{\theta} = 0$). We set no spin ($a_* = 0$) because the spin effects on the deformation only enter at higher orders. In such a limit, the line element of the KRZ metric reads

$$ds^2 = -\frac{N^2 - W^2}{K^2}\, dt^2 - 2\, Wr\; dt d\phi + K^2 r^2\; d\phi^2 + \Sigma r^2 d\theta^2 + \frac{\Sigma B^2}{N^2}\, dr^2 \tag{6.1}$$

where

$$N^2 = \left(1 - \frac{2M}{r}\right)\left[1 + \frac{8M^3\delta_1}{r^3}\right], \quad B = 1 + \frac{4M^2\delta_4}{r^2}, \tag{6.2}$$

$$\Sigma = 1, \quad K^2 = 1, \quad W = \frac{8M^3\delta_2}{r^3}. \tag{6.3}$$

The full metric without these assumptions can be seen in [17, 30]. From the metric, we follow the parametrized post-Einsteinian (ppE) formalism [34] to derive the parametrized GW phase, which can then be used to test the Kerr nature by constraining the deformations using the LVK data. In our analysis, we only recover the δ_1 and δ_2 parameters [30].

6.2.1 δ_1

From the normalization condition of the four-velocity, $u^\mu u_\mu = -1$,

$$g_{rr}\dot{r}^2 = -1 - g_{tt}\dot{t}^2 - g_{\phi\phi}\dot{\phi}^2 \equiv V_{\rm eff}, \tag{6.4}$$

introducing the effective potential $V_{\rm eff}$,

$$V_{\text{eff}} = -1 + E^2 + \frac{2M}{r} + \frac{L^2(2M-r)}{r^3} + \frac{8\delta_1 M^3(2M-r)}{r^4} + \frac{8\delta_1 L^2 M^3 (2M-r)}{r^6} + \mathcal{O}[\delta_1{}^2]. \tag{6.5}$$

where E and L are the specific energy and the specific angular momentum of a test-particle in circular orbit. They can be calculated by the condition $V_{\text{eff}} = dV_{\text{eff}}/dr = 0$. In the far-field limit, $L = r^2\dot{\phi} \to r^2\Omega$, where $\Omega = d\phi/dt$ is the angular velocity of the body as measured by a distant observer, we get a modified Kepler's law

$$\Omega^2 = \frac{M}{r^3}\left[1 + \frac{3M}{r} + \frac{9M^2}{r^2} - \frac{12M^2}{r^2}\delta_1 + \mathcal{O}\left(\delta_1{}^2, \frac{M^3}{r^3}\right)\right]. \tag{6.6}$$

The power of M in the δ_1 term represents the PN order [12], so we can say that δ_1 enters at 2PN. The orbital phase can be found by,

$$\phi(\nu) = \int^{\nu} \Omega\, dt = \phi_{\text{GR}}(\nu) - \frac{25}{4\eta}(2\pi m\nu)^{-1/3}\delta_1 + \mathcal{O}\left[\delta_1{}^2, (2\pi m\nu)^0\right], \tag{6.7}$$

Taking the Fourier transform,

$$\Psi_{\text{GW}}(f) = \Psi_{\text{GW}}^{\text{GR}}(f) - \frac{75}{8}u^{-1/3}\eta^{-4/5}\delta_1 + \mathcal{O}[\delta_1{}^2, u^0], \tag{6.8}$$

where $u = \eta^{3/5}\pi m f$. Comparing (6.8) to the ppE framework [34] and the LVK parametrization [21, 35],

$$\beta = -\frac{75}{8}\eta^{-4/5}\delta_1 = \frac{3}{128}\varphi_4\delta\varphi_4\eta^{-4/5}. \tag{6.9}$$

where, φ_4 is the PN phase at 2PN and has the form and $\delta\varphi_4$ is the deviation from GR [3, 5, 6] given as $\varphi_i \to (1+\delta\varphi_i)\varphi_i$ (i represents different PN phase). Finally, we find

$$\delta_1 = -\frac{1}{400}\varphi_4\delta\varphi_4. \tag{6.10}$$

which can be fit using the posteriors released by LVK. We analyze all the events present in the GWTC-2 catalogue and find a combined constraint by assuming each distribution follows a Gaussian distribution. For δ_1 the constraint for a 90% credible interval (CI) are,

$$\delta_1 = \begin{cases} -0.02 \pm 0.04 & (\texttt{IMRPhenomPv2}) \\ 0.02 \pm 0.05 & (\texttt{SEOBNRv4P}) \end{cases}.$$

IMRPhenomPv2 and SEOBNRv4P are the two waveform models used for the analysis by LVK. The constraints on each individual GWTC-2 events can be found in [30].

6.2.2 δ_2

In the case of δ_2 deformation parameter, one obtains,

$$\Omega^2 = \frac{M}{r^3}\left[1 + \frac{3M}{r} + \frac{9M^2}{r^2} + \frac{64M^{5/2}}{r^{5/2}}\delta_2 + \mathcal{O}\left(\delta_2{}^2, \frac{M^3}{r^3}\right)\right], \tag{6.11}$$

and

$$\delta_2 = -\left(\frac{7729\pi}{182784} - \frac{13\pi\eta}{2176}\right)\delta\varphi_5. \tag{6.12}$$

where φ_5 and $\delta\varphi_5$ are the phase and phase correction for 2.5PN. The combined constraints (90% CI) obtained on δ_2 are,

$$\delta_2 = \begin{cases} 0.01 \pm 0.02 & (\texttt{IMRPhenomPv2}) \\ 0.00 \pm 0.02 & (\texttt{SEOBNRv4P}) \end{cases}.$$

The constraints on individual GWTC-2 events can be found in [30].

6.3 Johannsen Metric

The Johannsen metric [15] is another popular axisymmetric parametrized metric for measuring deviations from Kerr spacetime. For equatorial geodesics, the metric reads,

$$ds^2 = -\frac{\Sigma\Delta}{B^2}\,dt^2 + \frac{\Sigma}{\Delta}\,dr^2 + \Sigma\left(d\theta^2 + d\phi^2\right), \tag{6.13}$$

where

$$\Sigma = r^2, \quad \Delta = r^2 - 2Mr, \quad B = r^2 A, \quad A = 1 + \alpha_{13}\left(\frac{M}{r}\right)^3. \tag{6.14}$$

Here, only one of the deformation parameters (α_{13}) of the metric is considered. The full metric with all other deformation parameters can be found in [15]. From here we obtain,

$$\Omega^2 = \frac{M}{r^3}\left[1 + \frac{3M}{r} + \frac{9M^2}{r^2} - \frac{3M^2\alpha_{13}}{r^2} + \mathcal{O}\left(\alpha_{13}{}^2, \frac{M^3}{r^3}\right)\right]. \tag{6.15}$$

and similar to δ_1 of the KRZ metric in the previous section we get a 2PN correction,

$$\alpha_{13} = -\frac{1}{100}\varphi_4\delta\varphi_4, \tag{6.16}$$

The combined constraints (90% CI) obtained for α_{13} are,

$$\alpha_{13} = \begin{cases} -0.10 \pm 0.16 & (\texttt{IMRPhenomPv2}) \\ 0.08 \pm 0.19 & (\texttt{SEOBNRv4P}) \end{cases}.$$

The constraints on individual GWTC-2 events can be found in [30].

6.4 Simpson-Visser Regular BH Metric

We also followed the same calculations for a regular BH metric coined by Simpson and Visser [23, 32]. This metric has a regularization parameter l which in our analysis we assume as a deformation and follow the analysis as in the previous sections. The metric reads,

$$ds^2 = -\left(1 - \frac{2M}{\Delta}\right)dt^2 + \left(1 - \frac{2M}{\Delta}\right)^{-1} dr^2 + \Delta^2\left(d\theta^2 + d\phi^2\right), \tag{6.17}$$

where $\Delta = \sqrt{r^2 + l^2}$. Here, $l > 0$ regularizes the singularity at the centre and different values of l gives rise to different scenarios. $l < 2M$ is a regular BH, $l = 2M$ gives rise to a one-way wormhole and $l > 2M$ forms a traversable two-way wormhole. The calculations following l as a deformation to Kerr leads,

$$\Omega^2 = \frac{M}{r^3}\left[1 + \frac{3M}{r} + \frac{9M^2}{r^2} - \frac{3M^2}{r^2}l^2 + \mathcal{O}\left(l^4, \frac{M^3}{r^3}\right)\right]. \tag{6.18}$$

and

$$l^2 = -\frac{1}{100}\varphi_4\delta\varphi_4. \tag{6.19}$$

The parameter l, similar to δ_1 of KRZ metric and α_{13} of Johannsen metric enters at 2PN. The constraints for l could not be combined as we constrained values of l^2, of which we have to reject the negative values so we provide there the strongest constraint (90% CI) we obtained with the event GW190707A,

$$l/M = \begin{cases} 0.44^{+0.24} & (\texttt{IMRPhenomPv2}) \\ 0.44^{+0.28} & (\texttt{SEOBNRv4P}) \end{cases} .$$

The constraints on rest of the GWTC-2 events can be found in [26]. We set the lower boundary for l/M to 0 because l is always positive by definition.

6.5 Conclusions

In the previous sections, we see the possibility of testing the Kerr nature of black holes by using the GW data against parametrized metrics. In all the cases we recover the Kerr metric in statistical limits. Such tests also act as agnostic tests of GR and also help in comparing the GR tests across multi-messenger observations as shown in Fig. 6.1.

6.5.1 Enhanced Constraints with Space-Based Detectors

The constraints on deviations from the Kerr metric obtained in this study are primarily based on the inspiral phase of binary black hole merger events observed by the

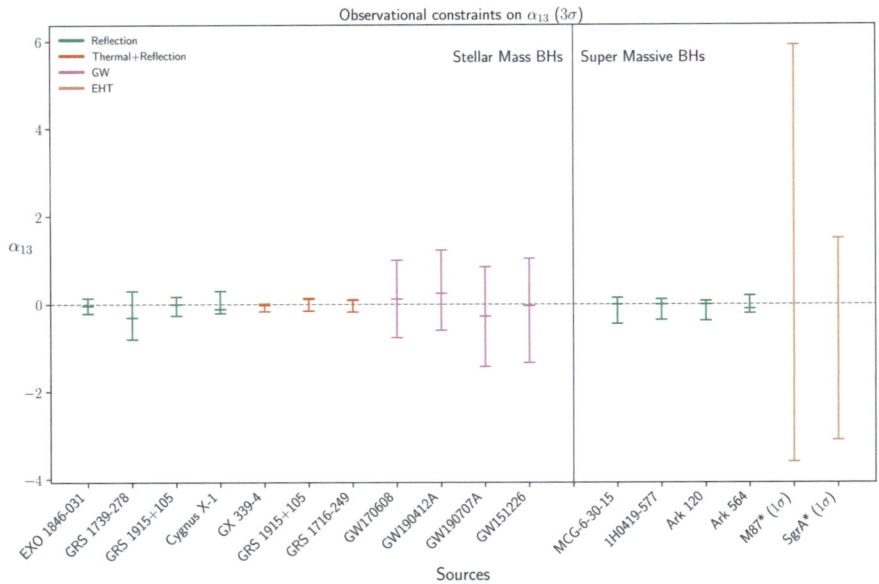

Fig. 6.1 Constraints for α_{13} parameter of the Johannsen metric [15] across X-ray, radio and GW measurements. Data from [8, 10, 25, 30]

LIGO-Virgo-KAGRA (LVK) collaboration [2, 4, 7]. These ground-based detectors are sensitive to GWs in the high-frequency band, capturing signals that last only a few seconds at most. Consequently, the inspiral phase observed is limited to the final moments before the merger, providing a relatively short window for extracting information about the spacetime geometry around the black holes.

However, space-based interferometers such as TianQin [22] and the Laser Interferometer Space Antenna (LISA) [27] are designed to operate in the low-frequency band of GWs, typically in the mHz range [28]. This sensitivity enables them to detect the inspiral phase of BBH mergers over much longer durations, ranging from hours to days or even weeks. The prolonged observation time will significantly enhance the signal-to-noise ratio and allow for much stringent tests of GR from the gravitational waveforms emitted during the inspiral phase [31]. TianQin might have a slight advantage as the maximum sensitivity of TianQin is in a higher frequency range than that of LISA and might provide better accuracy for such signals [18, 24].

References

1. B.P. Abbott et al., Observation of gravitational waves from a binary black hole merger. Phys. Rev. Lett. **116**(6), 061102 (2016)
2. B.P. Abbott et al., GWTC-1: a gravitational-wave transient catalog of compact binary mergers observed by LIGO and Virgo during the first and second observing runs. Phys. Rev. X **9**(3), 031040 (2019)
3. B.P. Abbott et al., Tests of general relativity with the binary black hole signals from the LIGO-Virgo catalog GWTC-1. Phys. Rev. D **100**(10), 104036 (2019)
4. R. Abbott et al., GWTC-2: compact binary coalescences observed by LIGO and Virgo during the first half of the third observing run. Phys. Rev. X **11**, 021053 (2021)
5. R. Abbott et al., Tests of general relativity with binary black holes from the second LIGO-Virgo gravitational-wave transient catalog. Phys. Rev. D **103**(12), 122002 (2021)
6. R. Abbott, et al., Tests of general relativity with GWTC-3 (2021), arXiv arXiv:2112.06861
7. R. Abbott et al., GWTC-3: compact binary coalescences observed by LIGO and Virgo during the second part of the third observing run. Phys. Rev. X **13**(4), 041039 (2023)
8. K. Akiyama, et al., First Sagittarius A* event horizon telescope results. VI. Testing the black hole metric. Astrophys. J. Lett. **930**, 2 (2022), L17
9. C. Bambi, Testing black hole candidates with electromagnetic radiation. Rev. Mod. Phys. **89**(2), 025001 (2017)
10. C. Bambi, Testing general relativity with black hole X-ray data: a progress report. Arab. J. Math. **11**(1), 81–90 (2022)
11. A. Cardenas-Avendano, S. Nampalliwar, N. Yunes, Gravitational-wave versus X-ray tests of strong-field gravity. Class. Quantum Grav. **37**(13), 135008 (2020)
12. Z. Carson, K. Yagi, Probing beyond-Kerr spacetimes with inspiral-ringdown corrections to gravitational waves. Phys. Rev. D **101**, 084050 (2020)
13. A. Einstein, Approximative integration of the field equations of gravitation. Sitzungsber. Preuss. Akad. Wiss. Berlin (Math. Phys.), 688–696 (1916)
14. M. Ghasemi-Nodehi, C. Bambi, Note on a new parametrization for testing the Kerr metric. Eur. Phys. J. C **76**, 1–7 (2016)
15. T. Johannsen, Regular black hole metric with three constants of motion. Phys. Rev. D **88**(4), 044002 (2013)
16. R.P. Kerr, Gravitational field of a spinning mass as an example of algebraically special metrics. Phys. Rev. Lett. **11**, 237–238 (1963)

17. R. Konoplya, L. Rezzolla, A. Zhidenko, General parametrization of axisymmetric black holes in metric theories of gravity. Phys. Rev. D **93**(6), 064015 (2016)
18. E.-K. Li, et al., Gravitational wave astronomy with TianQin
19. LIGO Scientific Collaboration, and Virgo Collaboration, Tests of general relativity with binary black hole signals from the LIGO-Virgo catalog GWTC-1 (2019), https://dcc-backup.ligo.org/LIGO-P1900087/public. Public data release, LIGO Document Number: LIGO-P1900087
20. LIGO Scientific Collaboration, and Virgo Collaboration, Tests of general relativity with binary black holes from the second LIGO-Virgo gravitational-wave transient catalog. Full Posterior Sample Data Release, September 2021
21. LIGO Scientific Collaboration, Virgo Collaboration, KAGRA Collaboration, LVK Algorithm Library—LALSuite, Free software (GPL) (2018)
22. J. Luo et al., TianQin: a space-borne gravitational wave detector. Class. Quant. Grav. **33**(3), 035010 (2016)
23. J. Mazza, E. Franzin, S. Liberati, A novel family of rotating black hole mimickers. J. Cosmol. Astropart. Phys. **2021**(04), 082 (2021)
24. J. Mei et al., The TianQin project: current progress on science and technology. PTEP **2021**(5), 05A107 (2021)
25. D. Psaltis, L. Medeiros, P. Christian, F. Özel, K. Akiyama, A. Alberdi, W. Alef, K. Asada, R. Azulay, D. Ball et al., Gravitational test beyond the first post-Newtonian order with the shadow of the M87 black hole. Phys. Rev. Lett. **125**(14), 141104 (2020)
26. S. Riaz, S. Shashank, R. Roy, A.B. Abdikamalov, D. Ayzenberg, C. Bambi, Z. Zhang, M. Zhou, Testing regular black holes with X-ray and GW data. JCAP **10**, 040 (2022)
27. P.A. Seoane et al., Astrophysics with the laser interferometer space antenna. Living Rev. Rel. **26**(1), 2 (2023)
28. N. Seto, K. Kyutoku, How many extragalactic stellar mass binary black holes will be detected by space gravitational-wave interferometers? Mon. Not. Roy. Astron. Soc. **514**(4), 4669–4675 (2022)
29. S. Shashank, Testing regularity of black holes with X-rays and gravitational waves, in *57th Rencontres de Moriond on Gravitation* (2023)
30. S. Shashank, C. Bambi, Constraining the Konoplya-Rezzolla-Zhidenko deformation parameters III: limits from stellar-mass black holes using gravitational-wave observations. Phys. Rev. D **105**(10), 104004 (2022)
31. C. Shi, M. Ji, J.-D. Zhang, J. Mei, Testing general relativity with TianQin: the prospect of using the inspiral signals of black hole binaries. Phys. Rev. D **108**(2), 024030 (2023)
32. A. Simpson, M. Visser, Black-bounce to traversable wormhole. JCAP **02**, 042 (2019)
33. A. Tripathi, Y. Zhang, A.B. Abdikamalov, D. Ayzenberg, C. Bambi, J. Jiang, H. Liu, M. Zhou, Testing general relativity with NuSTAR data of galactic black holes. Astrophys. J. **913**(2), 79 (2021)
34. N. Yunes, F. Pretorius, Fundamental theoretical bias in gravitational wave astrophysics and the parameterized post-Einsteinian framework. Phys. Rev. D **80**, 122003 (2009)
35. N. Yunes, K. Yagi, F. Pretorius, Theoretical physics implications of the binary black-hole mergers GW150914 and GW151226. Phys. Rev. D **94**(8), 084002 (2016)

Chapter 7
Joint Gravitational Wave Detection by TianQin and LISA

Alejandro Torres-Orjuela

Abstract We study the detection accuracy of double white dwarfs (DWDs), stellar-mass black hole binaries (SBHBs), light and heavy intermediate mass ratio inspirals (IMRIs), extreme mass ratio inspirals (EMRIs), massive black hole binaries (MBHBs), and the stochastic gravitational wave background (SGWB) of astronomical origin for TianQin, LISA, and joint detection. We use a Fisher matrix analysis and consider for each source the averaged detection accuracy over a realistic range of parameters. We find that on average TianQin obtains more accurate parameter estimation for DWDs and light IMRIs, LISA for heavy IMRIs, EMRIs, MBHBs, and the galactic foreground of the SGWB, and both contribute similarly to the detection of SBHBs and the extra-galactic SGWB. Nevertheless, for all sources joint detection allows setting tighter constraints on most parameters highlighting its importance for future detection.

7.1 Introduction

Gravitational wave (GW) detection has opened a new window to study compact objects [1–3] although it is currently restricted to the high-frequency band above 1 Hz covered by ground-based laser interferometry detectors LIGO, Virgo, and KAGRA [4–6] and the nHz band accessible through pulsar timing arrays [7–12]. The detectable spectrum will be expanded in the mid-2030s to the low-frequency (mHz) band by space-based laser interferometry detectors TianQin, LISA, and Taiji [13–17] while the intermediate band (dHz) will be opened by space-based laser interferometry detectors like DECIGO [18], ground-based and space-based atom interferometry detectors like AION, ZAIGA, and AEDGE [19–21] as well as lunar GW detectors like LGWA [22] in coming years.

A. Torres-Orjuela (✉)
Department of Physics, The University of Hong Kong, Pokfulam Road, Hong Kong
e-mail: atorreso@bimsa.cn

Beijing Institute of Mathematical Sciences and Applications, Beijing, China

J. Luo et al. (eds.), *Proceedings of the 7th International Workshop on the TianQin Science Mission; 25–26 April 2024, Hong Kong SAR, China*, Springer Proceedings in Physics 428, https://doi.org/10.1007/978-981-96-7803-7_7

TianQin and LISA are two space-based laser interferometer detectors with a similar design but differ in key aspects. Due to the differences in their design, TianQin and LISA can detect the same sources but obtain different accuracies. These sources include double white dwarfs (DWDs) which are the most abundant class of compact stellar mass binaries in our galaxy [23–25], stellar-mass black hole binaries (SBHBs) formed through co-evolving massive stars [26–28] or dynamical assembling in dense stellar systems [29–31], light intermediate mass ratio inspirals (IMRIs) formed by a stellar-mass black hole (BH) orbiting an intermediate-mass BH (IMBH) [32–34], heavy IMRIs formed by an IMBH orbiting a massive BH [35–37], extreme mass ratio inspirals (EMRIs) formed by a small compact object inspiraling into a MBH [38, 39], MBH binaries (MBHBs) residing in the center of galaxies [40–42], and the stochastic GW background (SGWB) resulting from the combination of a multitude of unresolved binaries [43–45].

We study the detection of astrophysical GW sources by TianQin, LISA, and joint detection using a Fisher matrix analysis [46–48]. The inverse of the Fisher Matrix $C = F^{-1}$ is a linear estimate of the measurement errors that asymptotes the true error in the limit of high signal-to-noise ratio. The detection accuracy of the joint detection is obtained by combining the results of the two independent detectors $F_{\rm Joint} = F_{\rm TianQin} + F_{\rm LISA}$ and then inverting $F_{\rm Joint}$. We focus on comparing the averaged detection accuracy of TianQin, LISA, and joint detection. However, the detection accuracy of a parameter can vary significantly depending on its specific value. Therefore, the results shown here shall only give an idea of the detection accuracy for the different detection scenarios and we refer the reader to [47] for more details. Moreover, when calculating the errors over the parameter range we only vary the considered parameter and fix the other parameters to fiducial values. The distributions used for the parameters are either uniform in linear space (lin), uniform in log-space (log), or uniform in cosine (cos).

7.2 Detection by TianQin, LISA, and Joint Detection

7.2.1 Double White Dwarfs

The parameters we consider for DWDs are the luminosity distance D_L, the chirp mass M_c, the inclination of the orbit relative to the line-of-sight ι, the orbital period of the binary P, and the sky localization Ω. We consider the sky localization error as a function of the elliptic colatitude θ and the longitude ϕ where one is fixed while the other is varied. The fiducial values, the parameter ranges considered and the distributions used, as well as the average errors are shown in Table 7.1. Note that the average errors for θ and ϕ refer to the sky localization error $\Delta\Omega$ as a function of these two parameters and are given in units of deg^2. We see on the left-hand side of Fig. 7.1 that TianQin constraints D_L, M_c, ι, and Ω_ϕ better than LISA while LISA can set tighter constraints than TianQin for P and Ω_θ. Therefore, we find that

Table 7.1 The fiducial values, the parameter ranges, and the averaged absolute errors in TianQin, LISA, and joint detection for DWDs

	D_L (kpc)	M_c ($M_\odot$)	ι (π)	P (s)	θ	ϕ
Fiducial	8.5	0.25	1/3	200	$-5.6°$	267°
Range	0.01–50	0.1–1.25	0–0.5	10–500	$-90°$–$+90°$	0°–360°
Distr.	log	lin	cos	lin	lin	lin
TianQin	1.2	4.6×10^{-4}	5.3×10^{-2}	8.9×10^{-4}	1.7	2.1×10^{-3}
LISA	1.8	6.1×10^{-4}	1.1×10^{-1}	2.5×10^{-4}	2.6×10^{-1}	1.3×10^{-2}
Joint	8.7×10^{-1}	3.6×10^{-4}	4.2×10^{-2}	2.4×10^{-4}	2.0×10^{-1}	1.5×10^{-4}

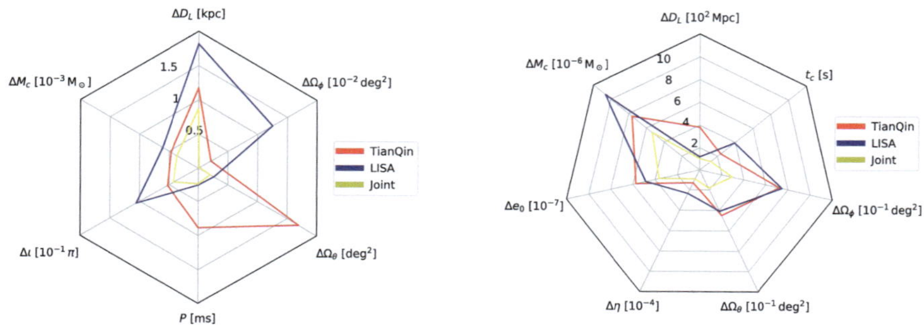

Fig. 7.1 The averaged absolute error for the parameters of DWDs (left) and SBHBs (right) measured by TianQin (red), LISA (blue), and joint detection (yellow)

although the detection of single parameters of a DWD might be dominated by one of the two detectors, joint detection yields significantly better results when studying all parameters combined. Note that as in real detection, all parameters are measured at the same time the improvements from joint detection are even more significant than they appear here. Last, we point out that the relatively big error in Ω_θ compared to Ω_ϕ for TianQin can be attributed to the particular choice of fiducial parameters and similar accuracies to those of Ω_ϕ can be expected for most sources.

7.2.2 *Stellar-Mass Black Hole Binaries*

For SBHBs, we consider D_L, M_c, the binary's eccentricity at 10 mHz e_0, the symmetric mass ratio η, Ω, and time to coalescence t_c. The errors are calculated over a parameter range where we only vary the considered parameter and fix the other parameters to fiducial values. The fiducial values, the parameter range, and the distributions applied, as well as the averaged absolute errors in TianQin, LISA, and joint detection can be found in Table 7.2. From the right side of Fig. 7.1 we see that M_c, η, and t_c are constrained better by TianQin while D_L and e_0 have smaller errors

Table 7.2 The fiducial values, the parameter ranges, and the averaged absolute errors in TianQin, LISA, and joint detection for SBHBs

	D_L (Mpc)	$\mathcal{M}_c$ ($M_\odot$)	e_0	η	θ	ϕ	t_c
Fiducial	200	50	0.05	0.125	+30°	180°	2.5 yr
Range	100–530	3.5–100	0.01–0.1	0.001–0.25	0°–+90°	0°–360°	1 yr-5 yr
Distr.	log	log	lin	log	cos	lin	log
TianQin	380	7.6×10^{-6}	5.7×10^{-7}	1.3×10^{-4}	4.5×10^{-1}	7.2×10^{-1}	2.3 s
LISA	110	1.1×10^{-5}	4.8×10^{-7}	2.4×10^{-4}	4.1×10^{-1}	7.5×10^{-1}	3.9 s
Joint	100	5.3×10^{-6}	3.6×10^{-7}	9.2×10^{-5}	1.8×10^{-1}	2.9×10^{-1}	1.2 s

Table 7.3 The fiducial values and the parameter ranges as well as the averaged detection errors in TianQin, LISA, and joint detection for light IMRIs

Parameter	D_L (Mpc)	M ($M_\odot$)	q	ι (π)	θ	ϕ
Fiducial	1	5000	100	1/3	0°	180°
Range	0.001–100	550–10^5	10–5000	0–0.5	0°–+90°	0°–360°
Distr.	log	log	log	lin	lin	lin
TianQin	2.6	35	5.1×10^{-1}	1.9×10^{-3}	2.4×10^{-4}	3.6×10^{-4}
LISA	15.5	169	3.9	1.2×10^{-2}	3.9×10^{-2}	1.7×10^{-2}
Joint	2.6	32	5.1×10^{-1}	1.9×10^{-3}	1.3×10^{-4}	2.4×10^{-4}

in LISA. Moreover, we see that Ω_θ and Ω_ϕ are constrained to a similar level by both detectors, and the small difference can be mainly attributed to the particular choice of fiducial values. In general, all parameters are constrained to a similar level by TianQin and LISA leading to a significant improvement in joint detection—in particular, for Ω, e_0, and $\mathcal{M}_c$.

7.2.3 Light Intermediate Mass Ratio Inspirals

We consider D_L, the total mass M, the mass ratio q, ι, and Ω as the parameters of light IMRIs, where the fiducial values, the parameter ranges used, and their distribution are shown in Table 7.3. The averaged absolute errors for TianQin, LISA, and joint detection can be found in the same table. The left-hand side of Fig. 7.2 shows the averaged detection error for light IMRIs. We see that the detection accuracy of TianQin is significantly better than the detection accuracy of LISA; often differing by almost an order of magnitude. Therefore, the accuracy of joint detection is dominated by TianQin but LISA can contribute to an improved detection of Ω and M.

Table 7.4 The fiducial values, the parameter ranges, and the averaged detection errors for heavy IMRIs by TianQin, LISA, and joint detection

Parameter	D_L (Mpc)	M ($M_\odot$)	q	ι (π)	θ	ϕ
Fiducial	30	10^6	100	1/3	0°	180°
Range	1–1000	10^5–$10^{6.2}$	10–5000	0–0.5	0°–+90°	0°–360°
Distr.	log	log	log	lin	lin	lin
TianQin	133	1.5×10^4	1.1	1.5×10^{-2}	5.6×10^{-3}	8.4×10^{-3}
LISA	32	5.8×10^3	3.1×10^{-1}	3.9×10^{-3}	3.7×10^{-3}	2.1×10^{-3}
Joint	31	5.3×10^3	3.0×10^{-1}	3.8×10^{-3}	1.1×10^{-3}	1.4×10^{-3}

7.2.4 *Heavy Intermediate Mass Ratio Inspirals*

For heavy IMRIs, we consider the same parameters as for a light IMRI, namely D_L, M, q, ι, and Ω and again compute the errors over a parameter range varying only the considered parameter while fixing the other parameters to fiducial values. In Table 7.4 we show the fiducial values, the range of the parameters, the distributions used for them, and the averaged detection error for TianQin, LISA, and joint detection. We, further, see on the right side of Fig. 7.2 that the detection accuracy in LISA is significantly better than TianQin's although the difference is less than one order of magnitude. Therefore, the constraints set by the joint detection are similar to those of LISA for D_L, M, q, and ι. For Ω the detection accuracy of the joint detection is also dominated by LISA but TianQin can considerably contribute to an improved measurement.

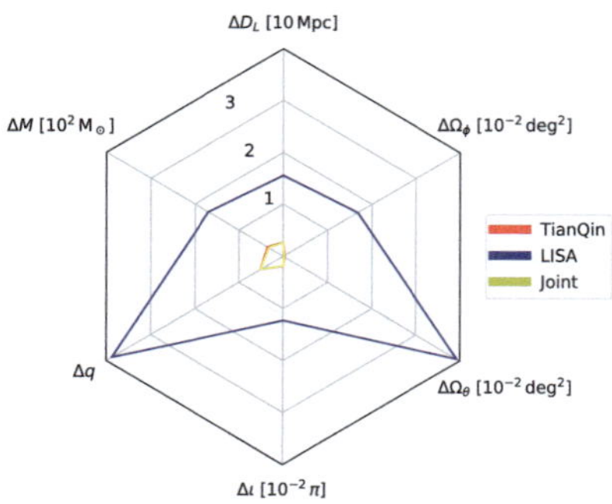

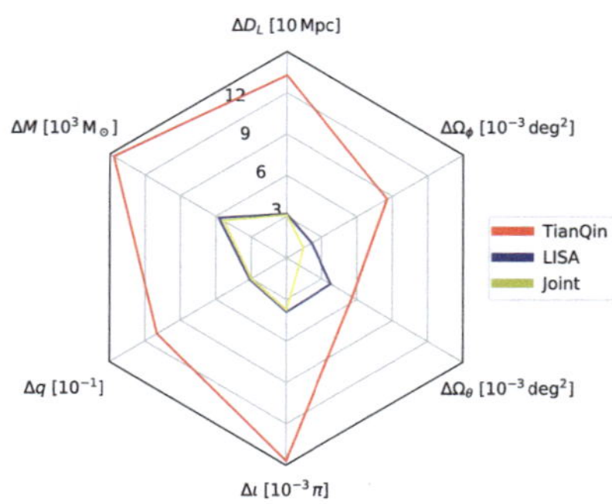

Fig. 7.2 The averaged absolute error for the parameters of light IMRIs (left) and heavy IMRIs (right) measured by TianQin (red), LISA (blue), and joint detection (yellow)

Table 7.5 The fiducial values, the parameter ranges, and the averaged absolute errors of the parameter in TianQin, LISA, and joint detection for EMRIs

Parameter	D_L (Gpc)	M_1 ($M_\odot$)	e_m	s	θ	ϕ
Fiducial	1	10^6	0.1	0.98	$+30°$	$0°$
Range	0.01–10	10^5–10^7	0.002–0.5	0.8–0.998	$-90°$–$+90°$	$0°$–$360°$
Distr.	log	log	lin	lin	cos	lin
TianQin	5.0×10^{-1}	6.5	7.9×10^{-5}	9.4×10^{-5}	1.8×10^{-1}	2.9×10^{-1}
LISA	1.6×10^{-1}	3.0	3.0×10^{-5}	3.0×10^{-5}	1.7×10^{-2}	1.7×10^{-2}
Joint	1.5×10^{-1}	2.7	2.8×10^{-5}	2.9×10^{-5}	1.7×10^{-2}	1.7×10^{-2}

7.2.5 *Extreme Mass Ratio Inspirals*

The parameters we study for an EMRI are ΔD_L, the mass of the MBH M_1, the eccentricity at merger e_m, the spin of the primary BH s, and Ω. We show their fiducial values, the parameter ranges considered and the distributions used, as well as the averaged detection errors by TianQin, LISA, and joint detection in Table 7.5. Moreover, we set the mass of the SBH orbiting the MBH to $10\,M_\odot$. The left-hand side of Fig. 7.3 shows that the detection accuracy of LISA is significantly better than the detection accuracy of TianQin but differs by less than one order of magnitude. Nevertheless, TianQin can contribute to slightly better constraints in the joint detection for D_L, M_1, e_m, and s. Only the accuracy of Ω in the joint detection is completely dominated by LISA which, however, can be partially attributed to the fiducial values assumed.

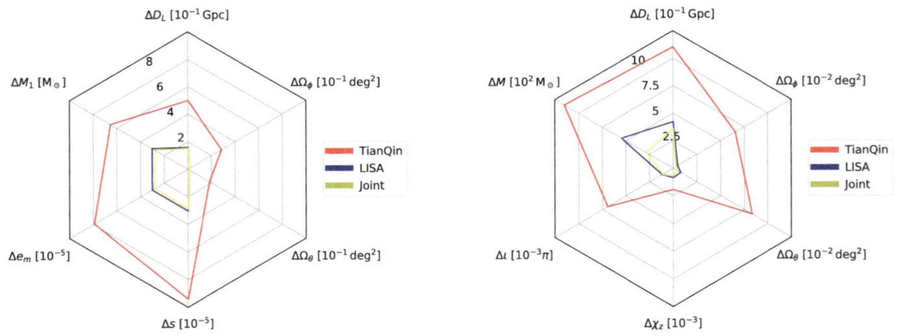

Fig. 7.3 The averaged absolute errors TianQin (red), LISA (blue), and joint detection (yellow) measure for the parameters of EMRI s(left) and MBHBs (right)

Table 7.6 The fiducial values and the parameter ranges as well as TianQin's, LISA's, and joint detection's averaged absolute errors for MBHBs

Parameter	D_L (Gpc)	M ($M_\odot$)	ι (π)	χ_z	θ	ϕ
Fiducial	10.5	4×10^6	1/3	0.5	$+45°$	$29.3°$
Range	0.5-50	10^4–10^7	0–1	0.04–0.96	$-90°$–$+90°$	$0°$–$360°$
Distr.	log	log	cos	lin	cos	lin
TianQin	1.1	1.2×10^3	7.0×10^{-3}	1.9×10^{-3}	8.3×10^{-2}	6.5×10^{-2}
LISA	4.2×10^{-1}	5.5×10^2	1.2×10^{-3}	8.6×10^{-4}	7.9×10^{-3}	4.6×10^{-3}
Joint	3.6×10^{-1}	2.8×10^2	1.1×10^{-3}	7.1×10^{-4}	4.2×10^{-3}	2.8×10^{-3}

7.2.6 *Massive Black Hole Binaries*

We consider D_L, M, ι, the effective spin of the BH along the angular momentum χ_z, and Ω as the parameters of MBHBs. We assume the components of the binary to be of equal mass and thus the effective spin refers to any of the two MBHs. Table 7.6 shows the fiducial values and the parameter ranges considered with the distributions applied, as well as the averaged detection error for TianQin, LISA, and joint detection. The errors are computed by varying the considered parameter over the parameter range while the other parameters are fixed to their fiducial values. On the right-hand side of Fig. 7.3, we see that LISA's detection error is smaller than TianQin's by a factor of two to three for most parameters. Only for Ω, the difference is more significant being of more than one order of magnitude. Therefore, the contribution by TianQin to the joint detection for Ω is small but the constraints on D_L and M are significantly improved in joint detection compared to LISA alone.

7.2.7 *Stochastic Gravitational Wave Background*

The energy spectrum density of the SGWB can be parametrized as

$$\Omega_{\rm GW} = \Omega_0 \left(\frac{f}{f_{\rm ref}}\right)^{\alpha_0} + \Omega_1 \left(\frac{f}{f_{\rm ref}}\right)^{\alpha_1} \left[1 + 0.75 \left(\frac{f}{f_{\rm ref}}\right)^{\Delta}\right]^{(\alpha_2-\alpha_1)/\Delta}, \tag{7.1}$$

where the first term represents the background formed by extra-galactic binaries [49–51] and the second term describes the foreground produced by Galactic DWDs [52]. Ω_0 is the amplitude level of the background at the reference frequency $f_{\rm ref} = 1\,$mHz and $\alpha_0 = 2/3$ is the spectral index for a background of binaries. Moreover, we set $\Omega_1 = 1 \times 10^{-6.94}$, $\alpha_1 = 3.64$, $\alpha_2 = -3.95$, and $\Delta = 0.75$. For the dimensionless energy density of the SGWB, we consider the range $10^{-12} \leq \Omega_{\rm GW} \leq 10^{-9}$ in accordance with the results of the LIGO-Virgo-KAGRA Collaboration [53].

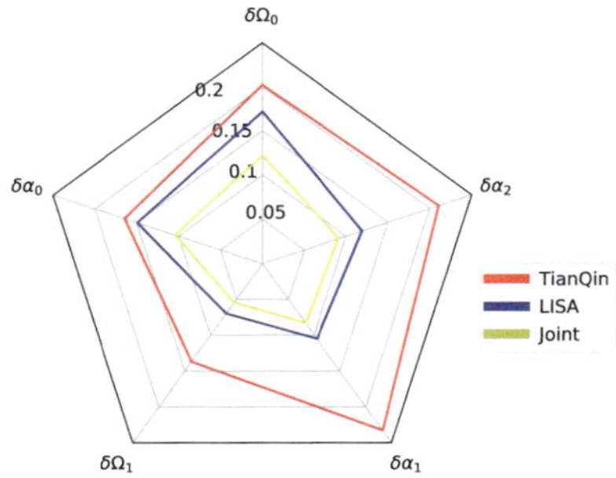

Fig. 7.4 The averaged relative error for the parameters of the extra-galactic background and the galactic foreground of the SGWB measured by TianQin (red), LISA (blue), and joint detection (yellow)

Table 7.7 The averaged detection errors in TianQin, LISA, and joint detection for the SGWB

Parameter	Ω_0	α_0	Ω_1	α_1	α_2
TianQin	2.0×10^{-1}	1.6×10^{-1}	1.4×10^{-1}	2.3×10^{-1}	2.1×10^{-1}
LISA	1.7×10^{-1}	1.5×10^{-1}	7.0×10^{-2}	1.1×10^{-1}	1.2×10^{-1}
Joint	1.2×10^{-1}	1.0×10^{-1}	5.5×10^{-2}	8.4×10^{-2}	9.1×10^{-2}

We see in Fig. 7.4 that the parameters of the background Ω_0 and α_0 are constrained slightly better by LISA than by TianQin. Therefore, both contribute significantly to the joint detection of these parameters. The detection of the foreground is dominated by LISA and thus joint detection of Ω_1, α_1, and α_2 only improves slightly by TianQin's contribution. For the detailed averaged relative errors see Table 7.7.

7.3 Summary

We study how accurately the parameters of DWDs, SBHBs, light and heavy IMRIS, EMRIs, MBHBs, and the SGWB formed by unresolved binaries are detected by TianQin, LISA, and joint detection. For each source, we consider a range of realistic parameters and compute the averaged detection error. We find that TianQin's detection accuracy is particularly good for DWDs and light IMRIs while LISA obtains tighter constraints for heavy IMRIs, EMRIs, MBHBs, and the galactic foreground of the SGWB. For SBHBs and the extra-galactic SGWB, TianQin and LISA obtain comparable results thus contributing similarly to joint detection. However, for all sources, joint detection yields improved detection accuracies showing the importance of coordinated detections. This study focuses on the averaged detection accuracy but detection errors can vary significantly depending on the specific value of the parameter. Therefore, we refer interested readers to [47] where the detection accuracy for TianQin, LISA, and joint detection are discussed in more detail. Although the discussion in [47] is much more extensive, the topic of joint detection requires further studies to understand the potential profit of coordinated detections by TianQin and LISA.

References

1. The LIGO Scientific Collaboration and the Virgo Collaboration, GWTC-1: A gravitational-wave transient catalog of compact binary mergers observed by LIGO and Virgo during the first and second observing runs. Phys. Rev. X **9**(3), 031040 (2019)
2. The LIGO Scientific Collaboration and the Virgo Collaboration, GWTC-2: Compact binary coalescences observed by LIGO and Virgo during the first half of the third observing run. Phys. Rev. X **11**(2), 021053 (2021)
3. The LIGO scientific collaboration, the Virgo collaboration, and the KAGRA collaboration. GWTC-3: Compact binary coalescences observed by LIGO and Virgo during the second part of the third observing run (2021), arXiv e-prints arXiv:2111.03606
4. LIGO Scientific Collaboration, Advanced LIGO. Class. Quantum Gravity **32**(7), 074001 (2015)
5. T. Accadia et al., Virgo: a laser interferometer to detect gravitational waves. J. Instrum. **7**(3), 3012 (2012)
6. Kagra Collaboration. KAGRA: 2.5 generation interferometric gravitational wave detector. Na. Astron. **3**, 35–40 (2019)
7. R.N. Manchester, IPTA, The international pulsar timing array. Class. Quantum Gravity **30**(22), 224010 (2013)
8. P.B. Demorest, R.D. Ferdman, M.E. Gonzalez, D. Nice, S. Ransom, I.H. Stairs, Z. Arzoumanian, A. Brazier, S. Burke-Spolaor, S.J. Chamberlin et al., Limits on the stochastic gravitational wave background from the north american nanohertz observatory for gravitational waves. Astrophys. J. **762**(2), 94 (2013)
9. R.N. Manchester, G. Hobbs, M. Bailes, W.A. Coles, W. van Straten, M.J. Keith, R.M. Shannon, N. Bhat, A. Brown, S.G. Burke-Spolaor et al., The Parkes pulsar timing array project. Publ. Astron. Soc. Aust. **30**, e017 (2013)
10. K.J. Lee, Prospects of gravitational wave detection using pulsar timing array for Chinese future telescopes, in *Frontiers in Radio Astronomy and FAST Early Sciences Symposium 2015*, ed. by L. Qain, D. Li, volume 502 of Astronomical Society of the Pacific Conference Series (2016), p. 19
11. G. Desvignes, R.N. Caballero, L. Lentati, J. Verbiest, D.J. Champion, B.W. Stappers, G.H. Janssen, P. Lazarus, S. Osłowski, S. Babak et al., High-precision timing of 42 millisecond pulsars with the European pulsar timing array. Mon. Not. R. Astron. Soc. **458**(3), 3341–3380 (2016)
12. B.C. Joshi, P. Arumugasamy, M. Bagchi, D. Bandyopadhyay, A. Basu, N.D. Batra, S. Bethapudi, A. Choudhary, K. De, L. Dey et al., Precision pulsar timing with the ORT and the GMRT and its applications in pulsar astrophysics. J. Astrophys. Astron. **39**(4), 51 (2018)
13. J. Luo, L.-S. Chen, H.-Z. Duan, Y.-G. Gong, H. Shoucun, J. Ji, Q. Liu, J. Mei, V. Milyukov, M. Sazhin et al., TianQin: a space-borne gravitational wave detector. Class. Quantum Gravity **33**(3), 035010 (2016)
14. P. Amaro-Seoane, H. Audley, S. Babak, J. Baker, E. Barausse, P. Bender, E. Berti, P. Binetruy, M. Born, D. Bortoluzzi, et al., Laser interferometer space antenna (2017), arXiv e-prints arXiv:1702.00786
15. X. Gong, Y.-K. Lau, S. Xu, P. Amaro-Seoane, S. Bai, X. Bian, Z. Cao, G. Chen, X. Chen, Y. Ding, et al. Descope of the ALIA mission, in *Journal of Physics Conference Series*, volume 610 of Journal of Physics Conference Series (2015), p. 012011
16. E.-K. Li, S. Liu, A. Torres-Orjuela, X. Chen, K. Inayoshi, L. Wang, Y.-M. Hu, P. Amaro-Seoane, A. Askar, C. Bambi, P.R. Capelo, H.-Y. Chen, A.J.K. Chua, E. Condés-Breña, L. Dai,

D. Das, A. Derdzinski, H.-M. Fan, M. Fujii, J. Gao, M. Garg, H. Ge, M. Giersz, S.-J. Huang, A. Hypki, Z.-C. Liang, B. Liu, D. Liu, M. Liu, Y. Liu, L. Mayer, N.R. Napolitano, P. Peng, Y. Shao, S. Shashank, R. Shen, H. Tagawa, A. Tanikawa, M. Toscani, V. Vázquez-Aceves, H.-T. Wang, S.-X. Yi, J.-D. Zhang, X.-T. Zhang, L. Zhu, L. Zwick, S. Huang, J. Mei, Y. Wang, Y. Xie, J. Zhang, J. Luo, Gravitational wave astronomy with TianQin (2024), arXiv e-prints arXiv:2409.19665
17. P.A. Seoane, M.A. Sedda, S. Babak, C.P.L. Berry, E. Berti, G. Bertone, D. Blas, T. Bogdanović, M. Bonetti, K. Breivik et al., The effect of mission duration on LISA science objectives. Gen. Relativ. Gravit. **54**(1), 3 (2022)
18. S. Kawamura, M. Ando, N. Seto, S. Sato, M. Musha, I. Kawano, J. Yokoyama, T. Tanaka, K. Ioka, T. Akutsu et al., Current status of space gravitational wave antenna DECIGO and B-DECIGO. Progr. Theor. Exp. Phys. **2021**(5), 05A105 (2021)
19. L. Badurina et al., AION: an atom interferometer observatory and network. JCAP **05**, 011 (2020)
20. M.-S. Zhan, J. Wang, W.-T. Ni, D.-F. Gao, G. Wang, L.-X. He, R.-B. Li, L. Zhou, X. Chen, J.-Q. Zhong et al., ZAIGA: Zhaoshan long-baseline atom interferometer gravitation antenna. Int. J. Mod. Phys. D **29**(4), 1940005 (2020)
21. Y.A. El-Neaj, C. Alpigiani, S. Amairi-Pyka, H. Araujo, A. Balaz, A. Bassi, L. Bathe-Peters, B. Battelier, A. Belic, E. Bentine et al., AEDGE: atomic experiment for dark matter and gravity exploration in space. EPJ Quantum Technol. **7**, 6 (2020)
22. P. Ajith, P.A. Seoane, M.A. Sedda, R. Arcodia, F. Badaracco, et al., The lunar gravitational-wave antenna: mission studies and science case (2024), arXiv e-prints, arXiv:2404.09181
23. G. Nelemans, L.R. Yungelson, S.F. Portegies Zwart, The gravitational wave signal from the Galactic disk population of binaries containing two compact objects. Astron. Astrophys. **375**, 890–898 (2001)
24. S. Yu, C.S. Jeffery, The gravitational wave signal from diverse populations of double white dwarf binaries in the Galaxy. Astron. Astrophys. **521**, A85 (2010)
25. K. Breivik, S. Coughlin, M. Zevin, C.L. Rodriguez, K. Kremer, C.S. Ye, J.J. Andrews, M. Kurkowski, M.C. Digman, S.L. Larson et al., COSMIC variance in binary population synthesis. Astrophys. J. **898**(1), 71 (2020)
26. D. Vanbeveren, The evolution of massive and very massive stars in clusters. New Astron. Rev. **53**(1–2), 27–35 (2009)
27. K. Belczynski, M. Dominik, T. Bulik, R. O'Shaughnessy, C. Fryer, D.E. Holz, The effect of metallicity on the detection prospects for gravitational waves. Astrophys. J. Lett. **715**(2), L138–L141 (2010)
28. M.U. Kruckow, T.M. Tauris, N. Langer, M. Kramer, R.G. Izzard, Progenitors of gravitational wave mergers: binary evolution with the stellar grid-based code COMBINE. Mon. Not. R. Astron. Soc. **481**(2), 1908–1949 (2018)
29. Z. Portegies, F. Simon, S.L.W. McMillan, Black hole mergers in the universe. Astrophys. J. Lett. **528**(1), L17–L20 (2000)
30. K. Gültekin, M. Coleman Miller, D.P. Hamilton, Growth of intermediate-mass black holes in globular clusters. Astrophys. J. **616**(1), 221–230 (2004)
31. H. Tagawa, Z. Haiman, B. Kocsis, Formation and evolution of compact-object binaries in AGN disks. Astrophys. J. **898**(1), 25 (2020)
32. C.M. Will, On the rate of detectability of intermediate-mass black hole binaries using LISA. Astrophys. J. **611**(2), 1080–1083 (2004)
33. P. Amaro-Seoane, J.R. Gair, M. Freitag, M. Coleman Miller, I. Mandel, C.J. Cutler, S. Babak, TOPICAL REVIEW: Intermediate and extreme mass-ratio inspirals—astrophysics, science applications and detection using LISA. Class. Quantum Gravity **24**(17), R113–R169 (2007)
34. M.A. Sedda, A. Seoane, X. Chen, Merging stellar and intermediate-mass black holes in dense clusters: implications for LIGO, LISA, and the next generation of gravitational wave detectors. Astron. Astrophys. **652**, A54 (2021)
35. P. Basu, S.K. Chakrabarti, Gravitational wave emission from a companion black hole in presence of an accretion disk around a super-massive Kerr black hole, in *American Institute of*

Physics Conference Series, ed. by S.K. Chakrabarti, A.S. Majumdar, volume 1053 of American Institute of Physics Conference Series (2008), pp. 33–37
36. M. Arca-Sedda, A. Gualandris, Gravitational wave sources from inspiralling globular clusters in the Galactic Centre and similar environments. Mon. Not. R. Astron. Soc. **477**(4), 4423–4442 (2018)
37. A. Derdzinski, D. D'Orazio, P. Duffell, Z. Haiman, A. MacFadyen, Evolution of gas disc-embedded intermediate mass ratio inspirals in the LISA band. Mon. Not. R. Astron. Soc. **501**(3), 3540–3557 (2021)
38. P. Amaro-Seoane, Relativistic dynamics and extreme mass ratio inspirals. Living Rev. Relativ. **21**(1), 4 (2018)
39. M. Mapelli, E. Ripamonti, A. Vecchio, A.W. Graham, A. Gualandris, A cosmological view of extreme mass-ratio inspirals in nuclear star clusters. Astron. Astrophys. **542**, A102 (2012)
40. S. Komossa, Observational evidence for supermassive black hole binaries, in *The Astrophysics of Gravitational Wave Sources*, ed. by J.M. Centrella. Volume 686 of American Institute of Physics Conference Series (2003), pp. 161–174
41. M. Milosavljević, D. Merritt, The final parsec problem, in *The Astrophysics of Gravitational Wave Sources*, ed. by J.M. Centrella. Volume 686 of American Institute of Physics Conference Series (2003), pp. 201–210
42. M. Milosavljević, D. Merritt, Long-term evolution of massive black hole binaries. Astrophys. J. **596**(2), 860–878 (2003)
43. B. Allen, J.D. Romano, Detecting a stochastic background of gravitational radiation: signal processing strategies and sensitivities. Phys. Rev. D **59**(10), 102001 (1999)
44. J.D. Romano, N.J. Cornish, Detection methods for stochastic gravitational-wave backgrounds: a unified treatment. Living Rev. Relativ. **20**(1), 2 (2017)
45. N. Christensen, Stochastic gravitational wave backgrounds. Rep. Progr. Phys. **82**(1), 016903 (2019)
46. D. Coe, Fisher matrices and confidence ellipses: a quick-start guide and software (2009), arXiv e-prints arXiv:0906.4123
47. A. Torres-Orjuela, S.-J. Huang, Z.-C. Liang, S. Liu, H.-T. Wang, C.-Q. Ye, H. Yi-Ming, J. Mei, Detection of astrophysical gravitational wave sources by TianQin and LISA. Sci. China Phys. Mech. Astron. **67**(5), 259511 (2024)
48. G. Wang, W.-B. Han, Alternative LISA-TAIJI networks: detectability of the isotropic stochastic gravitational wave background. Phys. Rev. D **104**(10), 104015 (2021)
49. A.J. Farmer, E.S. Phinney, The gravitational wave background from cosmological compact binaries. Mon. Not. R. Astron. Soc. **346**(4), 1197–1214 (2003)
50. T. Regimbau, The astrophysical gravitational wave stochastic background. Res. Astron. Astrophys. **11**(4), 369–390 (2011)
51. C.J. Moore, R.H. Cole, C. Berry, Gravitational-wave sensitivity curves. Class. Quantum Gravity **32**(1), 015014 (2015)
52. Z.-C. Liang, H. Yi-Ming, Y. Jiang, J. Cheng, J. Zhang, J. Mei, Science with the TianQin observatory: preliminary results on stochastic gravitational-wave background. Phys. Rev. D **105**(2), 022001 (2022)
53. LIGO Scientific Collaboration, Virgo Collaboration, KAGRA Collaboration, Upper limits on the isotropic gravitational-wave background from advanced LIGO and advanced Virgo's third observing run. Phys. Rev. D **104**(2), 022004 (2021)

Chapter 8
Astrophysics and Data Analysis for Space-Based Gravitational Wave Detectors

Yan Wang, Hong-Ming Jin, Wen-Fan Feng, Yao Fu, Tan Liu, Jie-Wen Chen, Soumya D. Mohanty, Yong Shao, and Bo Ma

Abstract The coming era of gravitational wave (GW) astronomy, enabled by space-based laser interferometric detectors, heralds unprecedented opportunities to study compact binaries, in particular double white dwarfs (DWDs), double neutron stars (DNSs), and binary black holes, through their GW and electromagnetic observations, providing important insights into binary evolution, NS physics, and the overarching

Y. Wang (✉) · H.-M. Jin · W.-F. Feng · Y. Fu · T. Liu · J.-W. Chen
National Gravitation Laboratory, MOE Key Laboratory of Fundamental Physical Quantities Measurements, Department of Astronomy and School of Physics, Huazhong University of Science and Technology, Wuhan, China
e-mail: ywang12@hust.edu.cn

W.-F. Feng
Kavli Institute for Astronomy and Astrophysics, Peking University, Beijing, China

T. Liu
School of Fundamental Physics and Mathematical Sciences, Hangzhou Institute for Advanced Study, UCAS, Hangzhou, China

University of Chinese Academy of Sciences, Beijing, China

J.-W. Chen
National Time Service Center, Chinese Academy of Sciences, Xi'an, China

Key Laboratory of Time and Frequency Primary Standards, Chinese Academy of Sciences, Xi'an, China

S. D. Mohanty
Department of Physics and Astronomy, University of Texas Rio Grande Valley, Brownsville, TX, USA

Department of Physics, IIT Hyderabad, Kandai, Telangana, India

Y. Shao
School of Astronomy and Space Science, Nanjing University, Nanjing, China

Key Laboratory of Modern Astronomy and Astrophysics, Nanjing University, Ministry of Education, Nanjing, China

B. Ma
School of Physics and Astronomy, Sun Yat-Sen University, Zhuhai, China

CSST Science Center in the Great Bay Area, Zhuhai, China

J. Luo et al. (eds.), *Proceedings of the 7th International Workshop on the TianQin Science Mission; 25–26 April 2024, Hong Kong SAR, China*, Springer Proceedings in Physics 428, https://doi.org/10.1007/978-981-96-7803-7_8

architecture of the Universe. In this paper, we provide an overview of our recent progress in the scientific objectives and data analysis relevant to these compact binaries within the scope of the space-based missions, emphasizing the multi-messenger observations of DWDs and DNSs in our Galaxy that synergize the capabilities of TianQin, LISA, and current and future optical and radio telescopes, and the promising multi-wavelength (dual-line) observations of the low-frequency GWs due to the binary motion of a DNS and the high-frequency GWs due to the spin of the aspherical NS component. We also briefly discuss the challenge of data analysis for the stellar-mass binary black holes for space-based detectors and present our preliminary solution, which involves both algorithmic innovations and hardware acceleration driven by graphics processing units (GPUs).

8.1 Introduction

White dwarfs (WDs) and neutron stars (NSs) are the final evolutionary stages of main sequence stars, each marking a distinct path in stellar evolution. Main sequence stars with masses $<8-10\ M_{\odot}$ typically evolve into WDs, while more massive stars, with initial masses ranging from about $8-25\ M_{\odot}$, end their lives as NSs following a core-collapse supernova explosion. It is estimated that more than 95% of the main sequence stars in our Galaxy will evolve into WDs, making them the most common stellar remnants. Neutron stars, though less numerous, play a crucial role in the dynamics of the Galaxy and are essential for understanding extreme physical conditions. The dense cores of WDs and NSs are supported against further gravitational collapse by electron degeneracy and neutron degeneracy pressure, respectively. When the mass of the core exceeds the Tolman–Oppenheimer–Volkoff limit ($\approx 2-3\ M_{\odot}$), no known forces are sufficient to prevent further collapse, and this will lead to the formation of a black hole (BH).

White dwarfs no longer produce energy through nuclear fusion reactions and gradually cool down over time, releasing their stored thermal energy and eventually fading. Neutron stars, on the other hand, are extremely dense and often exhibit strong magnetic fields and rapid rotation, resulting in observable phenomena such as pulsars. Both white dwarfs and neutron stars are essential for studying the final stages of stellar evolution, binary star interactions, and the overall structure of our Galaxy. Despite the challenge of observing these stellar remnants due to their low luminosities, large-scale surveys using ground and space telescopes in optical (such as ZTF, Gaia, etc.), radio (such as Parkes, FAST, etc.), and other bands have steadily increased the number of identified WD and NS candidates.

A significant fraction of main sequence stars, especially those with higher masses ($>1\ M_{\odot}$), form in binary systems [11, 27], which naturally leads to binaries that are composed of the compact stars. Among them, double white dwarfs (DWDs) and double neutron stars (DNSs) are particularly important because they provide valuable insights into the binary star evolution, the mechanisms underlying Type Ia supernovae and gamma-ray bursts. In the next decade, they are promising sources for space-based GW detectors, such as TianQin (TQ) [22] and the Laser Interferometer Space Antenna (LISA) [7].

In this paper, we summarize our recent progress for TQ and LISA in investigating the scientific objectives concerning the compact binaries, namely DWDs and DNSs, in our Galaxy, as well as a promising solution to the formidable data analysis challenge posed by SmBBHs. For the former, we simulate separate populations of DWDs and DNSs using binary population synthesis and analyze their detectability in GW and EM observations; and for the latter, we briefly present our method that addresses the challenge through both algorithmic innovations and hardware acceleration driven by graphics processing units (GPUs). For an overview of the astrophysical sources detectable by TianQin, one can refer to the review by Hu et al. in this volume, as well as to the recent TianQin Astrophysics White Paper [20].

8.2 Multi-messenger Detection of Galactic DWDs

In [18], we simulate two populations of DWDs originating from different star formation histories (referred to as Model 1 and Model 2) using the binary population synthesis code COSMIC (Compact Object Synthesis and Monte Carlo Investigation Code) [10]. Model 1 assumes a burst of star formation 13.7 billion years ago, while Model 2 assumes a constant star formation rate over 10 billion years for the thin disk, with bursts for the thick disk and bulge at 11 and 10 billion years ago, respectively.

Because of limited computational resources, we simulate only a subset of 1.7×10^8 DWDs. This fixed subset is scaled up to represent the entire population in the Galaxy according to the ratio between the mass of the zero-age main-sequence (ZAMS) stars used to create this fixed subset and the total mass of ZAMS of the different structures of our Galaxy [23]. These synthetic DWDs are then assigned a location in the Galaxy based on the mass density profile of the bulge, thin disk and thick disk described in [23].

Based on the antenna pattern functions and the noise power spectral densities of TQ and LISA, our study predicts the number of DWDs that can be detected by TQ and LISA individually and by their joint observation. We found that over the nominal mission durations, TQ, LISA, and their joint observation can detect approximately 5000 (10,000), 17,000 (33,000), and 18,000 (35,000) DWDs with signal-to-noise ratios (SNRs) greater than 7 in Model 1 (Model 2), respectively. Model 2 predicts roughly twice as many DWDs as Model 1, primarily because it generates more DWDs with frequencies in the millihertz range. Combined observations of TQ and LISA, shown in Fig. 8.1, can identify an additional 2000 DWDs compared to using LISA alone. Note that the squared SNR is proportional to the effective observation time. Due to the duty cycle of the detectors, which is 0.5 for TQ and 0.75 for LISA, the effective observation time is 2.5 years and 3 years for TQ and LISA, respectively.

In addition, an analysis based on the Fisher information matrix is performed to evaluate the estimation accuracy for the parameters in the GW signal, which is crucial for understanding the properties of these systems and their detectability by both

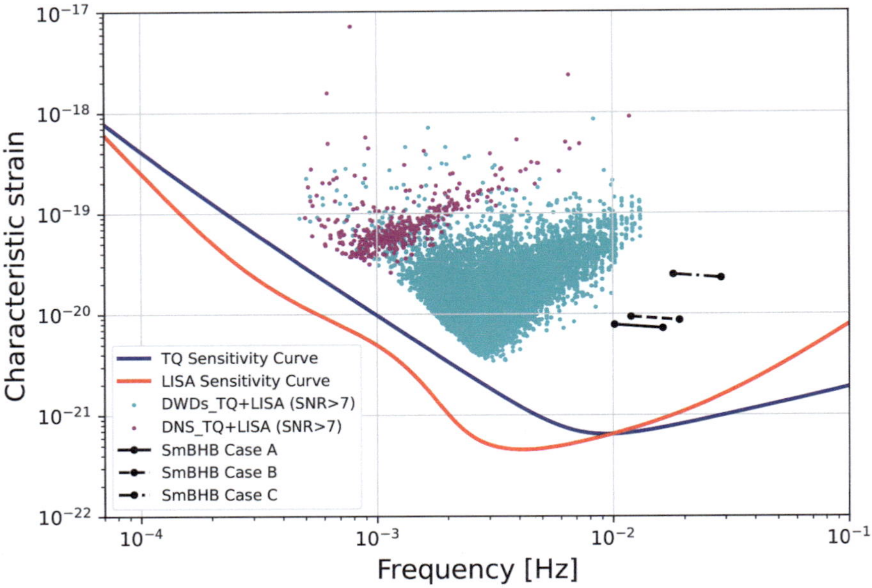

Fig. 8.1 The characteristic strain of detectable DWDs [18], DNSs [13], and SmBBHs [15]. Blue and red curves represent the sensitivity of TQ and LISA, respectively

GW and electromagnetic (EM) observations. TQ and LISA can individually detect approximately 220 (540) and 280 (730) DWDs, respectively, with relative errors in the chirp mass $\Delta\mathcal{M}/\mathcal{M}$ of <0.001 in Model 1 (Model 2). However, their joint observation increases the number to 390 (910), representing a 38% (24%) increase over LISA alone in Model 1 (Model 2). For relative errors in distance $\Delta\mathcal{D}/\mathcal{D}$, TQ and LISA can individually detect 920 (1770) and 1900 (3900) DWDs, respectively, with relative errors of <0.1 in Model 1 (Model 2). Their joint observation increases this number to 2200 (4600), an increase of $\approx$17% over LISA alone in both models. For error in sky location $\Delta\Omega$, TQ and LISA can individually detect 930 (1890) and 1900 (4010) DWDs, respectively, with errors of $<1\ deg^2$ in Model 1 (Model 2). Their joint observation increases this number to 2280 (4700), an increase of $\approx$ 20% over LISA alone in Model 1 (Model 2).

In the EM observation part, we generate the light curves for the simulated DWDs and investigate the expected number of detectable sources by Gaia and LSST. For the light curves, we use the PHOEBE (PHysics Of Eclipsing BinariEs) [26] with input parameters such as the effective temperatures and sizes of the stars, their distance from Earth, orbital period, orbital inclination, orbital eccentricity, etc. In our simulation, we also considered the secondary effects, such as the limb darkening and mutual irradiation, but we ignored the surface distortion due to tides and rotations. In the synthetic light curves, we convert the output flux from the simulation, measured in $W \cdot m^{-2}$, into apparent magnitudes for different bandpasses of Gaia (G-band) and LSST ($ugrizy$-bands). In this procedure, we have included the Gaussian noise rooted in the observation and the extinction effects due to dust in the interstellar environment [16, 21]. Unlike previous work [19], we used more realistic numbers of visits for the sky areas for Gaia and LSST and the errors in the apparent magnitude of the WDs [16, 17]. Our results indicate that Gaia and LSST can detect approximately 67 (186) and 273 (554) eclipsing DWDs, respectively, in Model 1 (Model 2). LSST shows a higher detection capability due to its deeper and more extensive sky survey. On the other hand, the number of detectable systems varies significantly between the two models, highlighting the importance of the underlying assumptions about star formation histories.

For the joint GW and EM observations, our simulation shows that TQ and Gaia can detect 2 (Model 1) and 9 (Model 2) eclipsing DWDs, while TQ and LSST can detect 7 (Model 1) and 11 (Model 2) eclipsing DWDs. In addition, LISA and Gaia can detect 4 (Model 1) and 29 (Model 2) eclipsing DWDs, while LISA and LSST can detect 15 (Model 1) and 47 (Model 2) eclipsing DWDs. Furthermore, 95% of the Galactic DWDs detectable by GW detectors are within 12 kpc, whereas 95% of the eclipsing DWDs observed by optical telescopes are within 4 kpc. This clearly demonstrates that GW detectors can provide a more comprehensive survey of DWDs in our Galaxy.

8.3 Multi-messenger and Multi-wavelength Detection of Galactic DNSs

Similarly, we investigate the feasibility and importance of multi-messenger observations of inspiraling double neutron stars in the Galactic disk, combining GWs and radio waves. In [13], we simulate the population of tight DNSs that emit GWs in the millihertz band following the model given in [8], and then study the potential of detection and parameter estimation of DNSs using TQ and LISA, as well as their joint observation (see Fig. 8.1). We also study the follow-up search for potential radio pulsars in the DNSs detectable by GWs, considering the characteristics of pulsars and the specifications of radio telescopes such as Parkes, FAST, SKA Phase 1, and SKA. The numbers of detectable pulsars and their timing precisions are reported.

In addition to the low-frequency GWs generated by the binary's orbital motion, the fast-spinning NS in the binary generates high-frequency GWs that can be detected by the current and next-generation ground-based detectors. However, the effects of spin-orbit coupling are enhanced in this tight binary, which will modify the waveforms of an isolated NS. In [14], we present such modified waveforms and demonstrate their impact on the detection and parameter estimation accuracy using the third-generation detector Cosmic Explorer [29].

With the two works above, it is natural to discuss the scientific benefits of an observing campaign combining the low- and high-frequency GWs (dual line) from such a DNS system in the mid-2030 when the space-based detector and the third-generation ground-based detectors are in simultaneous operation. In [12], using a combined four-year LISA and Cosmic Explorer observation, our study suggests that it is possible to determine the misalignment angle (the angle between the spin and total angular momentum) and the moment of inertia of the NS with sub-percentage and 5% accuracy, respectively.

An extended and more detailed version of this section is presented in a separate paper by *Feng et al.* in this volume.

8.4 Data Analysis for Stellar-Mass Binary Black Holes

The ground-based laser interferometric GW detector LIGO [1], in coordinated operation with Virgo [6] and KAGRA [28], has detected approximately one hundred GW events from compact binary coalescences during the observing runs O1–O3 [2–5]. The majority of these events are from stellar-mass binary black holes (SmBBHs). Months to years before their final mergers detected by the ground-based detectors, these binaries emit millihertz GWs during their inspiral phases, which are detectable by space-based detectors such as TQ and LISA.

The successful detections of SmBBHs in the low-frequency range will provide valuable insights into the formation channels of binary BHs and fundamental physics constraints, such as testing the BH no-hair theorem. However, the large number of wave cycles ($\mathcal{O}(10^5)$) accumulated in the millihertz range makes data analysis a formidable challenge for a matched filter-based or likelihood-based detection and characterization, which is extremely sensitive to the mismatches in the parameters of the signal and the templates. It has been shown that an astronomical number of templates ($10^{30}-10^{40}$) would be required to cover the signal parameter space for a blind search [25].

In [15], we introduce a hierarchical method to tackle this challenge, where a semi-coherent $\mathcal{F}$-statistic is calculated using varying numbers of frequency domain partitions at distinct stages. Our signal model adopts the restricted post-Newtonian waveform, where the 10 parameters $\{t_c, M_c, \delta_\mu, \theta, \phi, e_0, D_L, \psi, \iota, \phi_c\}$ representing the time of coalescence, the chirp mass, the dimensionless mass difference, the sky location angles (the ecliptic colatitude and longitude), the orbital eccentricity, the luminosity distance, the polarization angle, the angle between the line of sight and the angular momentum direction, and the phase at the coalescence. The first six parameters in this set are the intrinsic parameters, while the combination of the remaining parameters forms the set of the extrinsic parameters.

At each search stage, a chosen number of independent particle swarm optimization (PSO) [24] runs are used to globally optimize the semi-coherent $\mathcal{F}$-statistic in the intrinsic parameter space. A critical aspect of this method is the strategic partitioning of the parameter ranges for certain parameters at each stage in order to increase the convergence probability of the PSO and to prevent premature convergence to noise-induced secondary maxima in the semi-coherent $\mathcal{F}$-statistic. The parameter ranges for the semi-coherent searches are gradually narrowed down over the hierarchical stages, culminating in a final stage that searches for the global maximum of the fully-coherent $\mathcal{F}$-statistic.

As shown in Fig. 2 of [15], the main features of the three stages in our hierarchical pipeline are summarized as follows.

Stage 1: The search range for t_c is uniformly divided into 24 contiguous intervals. In each interval, 24 independent PSO runs are performed to search for the global maximum of the semi-coherent $\mathcal{F}$-statistic using $N_{\rm seg} = 16$ segments in the frequency domain. The number of iterations for each PSO run is 8000. The best solution from each interval is propagated to the next stage.

Stage 2: Further refine the search ranges for t_c and M_c based on the best solutions found in stage 1. The search ranges for all the other parameters, $N_{\rm seg}$, and the number of iterations for each PSO run remain the same as in stage 1. Each of the 24 intervals is searched with 8 independent PSO runs. The single best solution from these 192 PSO runs is propagated to the next stage.

Stage 3: Perform a fully-coherent $\mathcal{F}$-statistic search ($N_{\rm seg} = 1$) around the center formed by t_c, M_c, θ, ϕ, and e_0, as well as anther one with the same ranges except around $\pi - \theta$ due to the degeneracy in ecliptic colatitude. 48 PSO runs are implemented in each of the two search spaces with 15,000 iterations per run.

We test our method on three different SmBBH systems shown in Fig. 8.1. The injected signals have an SNR varying between $\approx$11 and $\approx$14 for a 2.5 years of a single LISA time delay interferometry (TDI) X-combination. The main difference in the injected values of the parameters is the chirp mass, which is used to test of our method at different points in the concerned search range. The luminosity distances in the three cases are adjusted to obtain the required SNRs. The noise realizations in the synthetic data are simulated based on the LISA instrumental noise [7, 9]. For the intrinsic parameters in the three cases, our method can estimate (i) the time of coalescence within $\approx$100 sec, (ii) the chirp mass with a relative error of $\lesssim$0.01%, (iii) the sky location within $\approx$0.2 deg^2, and (iv) orbital eccentricity at a fiducial signal frequency of 10 mHz with a relative error of $\lesssim$1%. However, for the dimensionless mass difference, the estimation error is significant even in the final stage. This is mainly due to the fact that the injected SmBBH signals in the current work are still in the relatively early inspiral phase, so that the post-Newtonian effect which is encoded in δ_μ is not strong enough.

The high precision measurement of the sky location for an SmBBH will allow us to search for its EM counterpart with ground- and space-based telescopes, so that one can perform multi-messenger investigation of the system. On the other hand, the prediction of the time of coalescence of an SmBHB well before its final mergers can inform the ground-based GW detectors, so that the detectors can be tuned to the optimized states for this source [30], maximizing the science payoff by join the observations of high- and low-frequency detectors.

Note that the hierarchical method described above involves several parallelizable steps. Each PSO run operates independently, and within each PSO run, the semi-coherent $\mathcal{F}$-statistic for each particle can be computed in parallel. Additionally, computations for each frequency in the frequency-domain response can be parallelized.

This multi-layer parallelization is implemented using a nested set of hardware elements: cluster nodes at the top level, CPU cores at the intermediate level, and GPUs at the deepest level. Currently, this parallelization scheme enables the hierarchical search to complete the analysis of a single data realization in approximately 2.5 d using 8 NVIDIA A100 GPUs and 48 CPU cores on a single compute node.

8.5 Conclusions

This paper provides a brief overview of our recent progress in astrophysics and data analysis for space-based GW detectors, with a special focus on TianQin. We have investigated the feasibility and advantages of multi-messenger and multi-wavelength observations for double white dwarfs and double neutron stars in our Galaxy. In addition, we proposed a preliminary solution to the data analysis challenges associated with stellar-mass binary black holes. Extensions of these works in various aspects are currently under investigation.

Acknowledgements Y.W. gratefully acknowledges support from the National Key Research and Development Program of China (Nos. 2023YFC2206702 and 2022YFC2205201), the National Natural Science Foundation of China (NSFC) under Grants No. 11973024, Major Science and Technology Program of Xinjiang Uygur Autonomous Region (No. 2022A03013-4), and Guangdong Major Project of Basic and Applied Basic Research (Grant No. 2019B030302001). T. L. is supported by NSFC Grant No. 12003008. J.-W. C. acknowledges the support from China Postdoctoral Science Foundation under Grant No. 2021M691146. S.D.M is supported by U.S. National Science Foundation (NSF) grant PHY-2207935. Y.S. acknowledges support from the National Key Research and Development Program of China (Grant Nos 2021YFA0718500 and 2023YFA1607902). B.M. acknowledges support from NSFC (12073092, 12103097).

References

1. J. Aasi, B. Abbott, R. Abbott, T. Abbott, M. Abernathy et al., Advanced ligo. Class. Quantum Grav. **32**(7), 074001 (2015)
2. B. Abbott, R. Abbott, T. Abbott, S. Abraham, F. Acernese et al., Gwtc-1: a gravitational-wave transient catalog of compact binary mergers observed by ligo and virgo during the first and second observing runs. Phys. Rev. X **9**(3), 031040 (2019)
3. B.P. Abbott, R. Abbott, T. Abbott, M. Abernathy, F. Acernese et al., Binary black hole mergers in the first advanced ligo observing run. Phys. Rev. X **6**(4), 041015 (2016)
4. R. Abbott, T. Abbott, S. Abraham, F. Acernese, K. Ackley et al., Gwtc-2: compact binary coalescences observed by ligo and virgo during the first half of the third observing run. Phys. Rev. X **11**(2), 021053 (2021)
5. R. Abbott, T. Abbott, F. Acernese, K. Ackley, C. Adams et al., Gwtc-3: compact binary coalescences observed by ligo and virgo during the second part of the third observing run. Phys. Rev. X **13**(4), 041039 (2023)
6. F. Acernese, M. Agathos, K. Agatsuma, D. Aisa, N. Allemandou, A. Allocca, J. Amarni, P. Astone, G. Balestri, G. Ballardin et al., Advanced virgo: a second-generation interferometric gravitational wave detector. Class. Quantum Grav. **32**(2), 024001 (2015)

7. P. Amaro-Seoane, H. Audley, S. Babak, et al., Laser interferometer space antenna (2017), arXiv e-prints arXiv:1702.00786
8. J.J. Andrews, K. Breivik, C. Pankow, D.J. D'Orazio, M. Safarzadeh, Lisa and the existence of a fast-merging double neutron star formation channel. Astrophys. J. Lett. **892**(1), L9 (2020)
9. J.W. Armstrong, F.B. Estabrook, M. Tinto, Time-delay interferometry for space-based gravitational wave searches. Astrophys. J. **527**(2), 814 (1999)
10. K. Breivik, S. Coughlin, M. Zevin, C.L. Rodriguez, K. Kremer, C.S. Ye, J.J. Andrews, M. Kurkowski, M.C. Digman, S.L. Larson, F.A. Rasio, Cosmic variance in binary population synthesis. Astrophys. J. **898**(1), 71 (2020)
11. K. El-Badry, H.-W. Rix, T.M. Heintz, A million binaries from gaia edr3: sample selection and validation of gaia parallax uncertainties. Month. Not. Royal Astron. Soc. **506**(2), 2269–2295 (2021)
12. W.-F. Feng, J.-W. Chen, T. Liu, Y. Wang, S.D. Mohanty, Inferring binary parameters with dual-line gravitational wave detection from tight inspiraling double neutron stars. Phys. Rev. D **109**, 043033 (2024)
13. W.-F. Feng, J.-W. Chen, Y. Wang, S.D. Mohanty, Y. Shao, Multimessenger observations of double neutron stars in the galactic disk with gravitational and radio waves. Phys. Rev. D **107**, 103035 (2023)
14. W.-F. Feng, T. Liu, J.-W. Chen, Y. Wang, S.D. Mohanty, Effects of spin-orbit coupling on gravitational waveforms from a triaxial nonaligned neutron star in a binary system. Phys. Rev. D **108**, 063035 (2023)
15. Y. Fu, Y. Wang, S.D. Mohanty, Hierarchical search method for gravitational waves from stellar-mass binary black holes in noisy space-based detector data. Phs. Rev. **111**(4), 043026 (2025), arXiv e-prints arXiv:2407.10797. https://doi.org/10.1103/PhysRevD.111.043026
16. Gaia Collaboration, T. Prusti, J.H.J. de Bruijne, et al., The Gaia mission. Astron. Astrophys. **595**, A1 (2016)
17. Ž. Ivezić, S.M. Kahn, J.A. Tyson et al., LSST: from science drivers to reference design and anticipated data products. Astrophys. J. **873**(2), 111 (2019)
18. H.-M. Jin, B. Ma, Y. Shao, Y. Wang, Detecting eclipsing double white dwarfs with electromagnetic and gravitational waves. Month. Not. Royal Astron. Soci. **536**(3), 2770–2782 (2025), arXiv eprint arXiv:2406.16474. https://doi.org/10.1093/mnras/stae2696
19. V. Korol, E.M. Rossi, P.J. Groot, G. Nelemans, S. Toonen, A. Brown, Prospects for detection of detached double white dwarf binaries with gaia, lsst and lisa. Month. Not. Royal Astron. Soc. **470**(2), 1894–1910 (2017)
20. E.-K. Li, S. Liu, A. Torres-Orjuela, X. Chen, K. Inayoshi, L. Wang, Y.-M. Hu, P. Amaro-Seoane, A. Askar, C. Bambi, P.R. Capelo, H.-Y. Chen, A.J.K. Chua, E. Condés-Breña, L. Dai, D. Das, A. Derdzinski, H.-M. Fan, M. Fujii, J. Gao, M. Garg, H. Ge, M. Giersz, S.-J. Giersz, A. Hypki, Z.-C. Liang, B. Liu, D. Liu, M. Liu, Y. Liu, L. Mayer, N.R. Napolitano, P. Peng, Y. Shao, S. Shashank, R. Shen, H. Tagawa, A. Tanikawa, M. Toscani, V. Vázquez-Aceves, H.-T. Wang, H. Wang, S.-X. Yi, J.-D. Zhang, X.-T. Zhang, L. Zhu, L. Zwick, S. Huang, J. Mei, Y. Wang, Y. Xie, J. Zhang, J. Luo, Gravitational wave astronomy with TianQin. Rep. Prog. Phys. **88**(5), 056901 (2025), arXiv e-printsarXiv:2409.19665. https://doi.org/10.1088/1361-6633/adc9be
21. LSST Science Collaboration, P.A. Abell, J. Allison, S.F. Anderson, et al., LSST science book, version 2.0 (2009), arXiv e-prints arXiv:0912.0201
22. J. Luo, L.-S. Chen, H.-Z. Duan, Y.-G. Gong, S. Hu, J. Ji, Q. Liu, J. Mei, V. Milyukov, M. Sazhin, C.-G. Shao, V.T. Toth, H.-B. Tu, Y. Wang, Y. Wang, H.-C. Yeh, M.-S. Zhan, Y. Zhang, V. Zharov, Z.-B. Zhou, Tianqin: a space-borne gravitational wave detector. Class. Quantum Grav. **33**(3), 035010 (2016)
23. P.J. McMillan, Mass models of the milky way. Month. Not. Royal Astron. Soc. **414**(3), 2446–2457 (2011)
24. S. Mohanty, *Swarm Intelligence Methods for Statistical Regression* (CRC Press, 2018)
25. C.J. Moore, D. Gerosa, A. Klein, Are stellar-mass black-hole binaries too quiet for lisa? Month. Not. Royal Astron. Soc.: Lett. **488**(1), L94–L98 (2019)

26. A. Prša, *Modeling and Analysis of Eclipsing Binary Stars* (IOP Publishing, 2018), pp. 2514–3433
27. H. Sana, S.E. de Mink, A. de Koter, N. Langer, C.J. Evans, M. Gieles, E. Gosset, R.G. Izzard, J.B. Le Bouquin, F. Schneider, Binary interaction dominates the evolution of massive stars. Science **337**(6093), 444 (2012)
28. K. Somiya, Detector configuration of kagra-the japanese cryogenic gravitational-wave detector. Class. Quantum Grav. **29**(12), 124007 (2012)
29. V. Srivastava, D. Davis, K. Kuns, P. Landry, S. Ballmer, M. Evans, E.D. Hall, J. Read, B.S. Sathyaprakash, Science-driven tunable design of cosmic explorer detectors. Astrophys. J. **931**(1), 22 (2022)
30. R. Tso, D. Gerosa, Y. Chen, Optimizing ligo with lisa forewarnings to improve black-hole spectroscopy. Phys. Rev. D **99**, 124043 (2019)

Chapter 9
Parameter Estimation of Stellar Mass Binary Black Holes in the Network of TianQin and LISA

Xiangyu Lyu, Enkun Li, and Yi-Ming Hu

Abstract We present a Bayesian parameter estimation progress to infer the stellar mass binary black hole properties by TianQin, LISA, and joint observation(TianQin+LISA). Two typical stellar-mass Black Hole Binary systems, GW150914 and GW190521 are chosen as the fiducial sources. In this work, we establish the ability of TianQin to infer the parameters of those systems and first apply the full frequency response in TianQin's data analysis. We obtain the parameter estimation results and explain the correlation between different parameters. The results show that TianQin+LISA could marginally increase the parameter estimation precision and narrow the 1σ area compared with TianQin and LISA individual observations. We finally demonstrate the importance of considering the effect of spin when the binaries have a non-zero component spin and great deviation will appear especially on mass, coalescence time and sky location.

9.1 Background

In 2015, the detection of a gravitational wave (GW) event [1] brought scientists a brand-new way of exploring the universe, and such achievement enabled three physicists in the GW detection team LIGO-Virgo-KAGRA(LVK) to share the 2017 Nobel Prize in Physics. In the meanwhile, the detection of GW provides a new method of observing the black hole(BH), which allows physicists to directly measure the physical properties of BH and invert their parameter information for the first time. Compared with the previous electromagnetic observations, which had various limitations and made it impossible for astronomers to accurately describe the evolution progress of binary stars, the direct measurement of BH physical parameters through

X. Lyu · E. Li · Y.-M. Hu (✉)
MOE key Laboratory of TianQin Mission, TianQin Research Center for Gravitational Physics and Astronomy, Frontiers Science Center for TianQin, Gravitational Wave Research Center for CNSA, Sun Yat-sen university (Zhuhai Campus), Zhuhai, China
e-mail: huyiming@mail.sysu.edu.cn

X. Lyu
e-mail: lvxy55@mail.sysu.edu.cn

J. Luo et al. (eds.), *Proceedings of the 7th International Workshop on the TianQin Science Mission; 25–26 April 2024, Hong Kong SAR, China*, Springer Proceedings in Physics 428, https://doi.org/10.1007/978-981-96-7803-7_9

GW detection has enabled astronomers to provide evidence for the study of evolution mechanisms from the observation data for the first time.

Based on the evolution stages of stellar-mass binary black holes (sBBHs), their waveform can be categorized into inspiral, merger, and ringdown phases, spanning frequencies from millihertz in the early stages to hundreds of hertz during the merger. And a fraction of the observed sBBH systems [2] can be detected by both LISA [3] and TianQin [4], which also allows for joint observation (TianQin + LISA) and it has been proved that space-borne GW detectors have significant ability to extract the sBBH systems' physics parameter information. Space-borne GW detectors could typically observe the early inspiral stage sBBH GW signals that fall outside the sensitive frequency bands of the ground-based GW detectors. During this stage, the orbital eccentricity has not yet significantly lost [5]. Moreover, in the space sensitive band, sBBH systems can persist from months to years, enabling precise characterization of the GW phase through the accumulation of observation data. For these reasons, space-borne GW detectors could not only achieve high parameter estimation precision for source mass [6–9], eccentricity [8, 10, 11], and spin [12], but accurately locate the host galaxies of the sources [8].

Furthermore, for space-borne GW detectors, the conventional template-bank approaches adopted by the ground-base GW detectors have been demonstrated to be quite expensive for sBBH signal searching [13] and there are some researches which promoted different method to achieve the sBBH signal search [14–16].

Based on the above research background, the objectives of this paper are first to develop a basic data processing tool for TianQin targeting stellar-mass BBHs and then, under the context of multi-detector joint observations, to extract information from simulated observational data, present the parameter estimation results.

9.2 Method

The detection of the GW signals allows us to directly probe the sources' physical information. With the detection data, the space-borne GW detectors have been proven they could give a accurate parameter estimation precision. To achieve this aim, in this paper, for the first time in this research area, we establish the full-frequency response of TianQin and then generate the mock data. Besides, we have developed the parameter estimation tools, enabling us to successfully derive the parameters' posterior distributions from the observation data.

9.2.1 sBBH Waveform

The stellar-mass binary black hole systems reside in the inspiral phase within the sensitive frequency band of space-based gravitational wave detectors, and the gravitational waves emitted during this phase can be accurately described by the IMRPhenomD waveform. Furthermore, to facilitate a reasonable comparison with the work presented in [17, 18], we have also adopted the same waveform to generate our signals. Consequently, in terms of waveform selection, the IMRPhenomD waveform is employed.

In this paper, we consider a simplified model for binary black holes, where the parameters are solely comprised of $\theta = (D_L, M_c, \eta, \iota, \lambda, \beta, \psi, t_c, \phi_c, \chi_1, \chi_2)$. Here, D_L represents the luminosity distance, ι is the inclination angle between the binary's orbital angular momentum and the line of sight, M_c and η are mass parameters given by $M_c = (m_1 m_2)^{3/5} / (m_1 + m_2)^{1/5}$ and $\eta = m_1 m_2 / (m_1 + m_2)^2$, respectively, where m_1 and m_2 are the masses of the primary and secondary black holes. Additionally, λ and β represent the spatial coordinates (longitude and latitude), while ψ, t_c, and ϕ_c denote the polarization angle, coalescence time, and coalescence phase, respectively. Lastly, χ_1 and χ_2 are the spin parameters of the two black holes, assuming aligned spins, i.e., their spin directions are parallel to the orbital angular momentum, and eccentricity and precession effects are neglected.

9.2.2 Response Model

In the previous, science case studies for LISA [6, 7] and TianQin [8, 19] utilize simplification response model and it wouldn't significantly alter signal-to-noise ratio (SNR) while greatly complicating the calculations. However, in reality, since the laser phase noise can be orders of magnitude higher than the GW signals, one has to rely on the Time Delay Interferometry(TDI) combination of channels to cancel out the laser phase noise [20–23]. In particular, we adopt the A, E, T [24] combination, with A and E reflecting the two GW polarizations and the T channel is a noise-monitoring channel. In this work, we utilized the analytical full frequency response described in [17, 25] to ensure speed.

The TDI response of a space-based gravitational wave detector can be initially represented by the observable y_{slr} of a single detector link. The time-domain response of a single laser link is given as follows,

$$y_{slr} = \frac{1}{2} \frac{\hat{n}_l \otimes \hat{n}_l}{1 - \hat{k} \cdot \hat{n}_l} : \left[h\left(t - L - \hat{k} \cdot \mathbf{p}_s\right) - h\left(t - \hat{k} \cdot \mathbf{p}_r\right) \right], \tag{9.1}$$

where $\hat{k}$ is the direction of gravitational wave propagation, $\hat{n}_l$ is the unit vector from satellite r to satellite s, h is the gravitational wave tensor, and L is the detector arm length.

Based on the time-domain response, the frequency-domain response is constructed as:

$$
\begin{aligned}
T_{slr}(f) &= G_{slr}(f, t(f)) \\
&= i\pi f L/c \,\mathrm{sinc}\left[\pi f L(1 - \hat{k}\cdot\hat{n}_l)/c\right] \\
&\cdot \exp\left[i\pi f(L + \hat{k}\cdot(\mathbf{p}_r + \mathbf{p}_s))/c\right]\hat{n}_l \cdot \mathbf{P}\cdot \hat{n}_l,
\end{aligned}
\tag{9.2}
$$

where $\mathbf{p}_r$ and $\mathbf{p}_s$ are the position vectors of the detectors in the coordinate system, $\mathbf{P}$ is the polarization tensor, and $t(f) = -\frac{1}{2\pi}\frac{d\Psi}{df}$ represents the time-frequency relationship. This response remains valid in the high-frequency region.

For the TDI channels, it is assumed that the detector arm lengths do not vary with time, and the response signals of channels A, E, and T in the TDI combination can be composed of Michelson channels X, Y, and Z, where $y_{(sr,nL)} = y_{sr}(t - nL)$. The combinations are given by:

$$
X = y_{31} + y_{(13,L)} + (y_{(12,L)} + y_{21})_{,2L} - (y_{31} + y_{(13,L)})_{,2L} - (y_{21} + y_{(12,L)}).
\tag{9.3}
$$

The corresponding channel response functions are:

$$
A = \frac{1}{\sqrt{2}}(Z - X), \quad E = \frac{1}{\sqrt{6}}(X - 2Y + Z), \quad Z = \frac{1}{\sqrt{3}}(X + Y + Z).
\tag{9.4}
$$

The frequency-domain response signals for A, E, and T are:

$$
\tilde{y}_{slr} = T_i(f)\tilde{h}(f), \quad i = A, E, T.
\tag{9.5}
$$

9.3 Results

9.3.1 Mock Data Generation

In this paper, we utilize the GW150914-like and GW190521-like sources as our fiducial systems and generate the TDI response signal under TianQin and LISA.Furthermore, the noise realization was assumed to be zero in the simulated observation data for TianQin and LISA (Fig. 9.1).

The left figure shows the response signals in the A channel for the GW150914-like source in TianQin and LISA, along with the Amplitude Spectrum Density (ASD) of the channel noise. The 03A,69A,1215A,1821A labels in both subfigure represent the observation data in A channel of TianQin under 3+3 model. Especially, the '03A' demonstrates that the detector's telescopes operate in the first three month ($0 \sim 3$mon) and get the observation data in A channel. The right figure presents the same for a GW190521-like source in TianQin and LISA, also including the channel noise ASD.

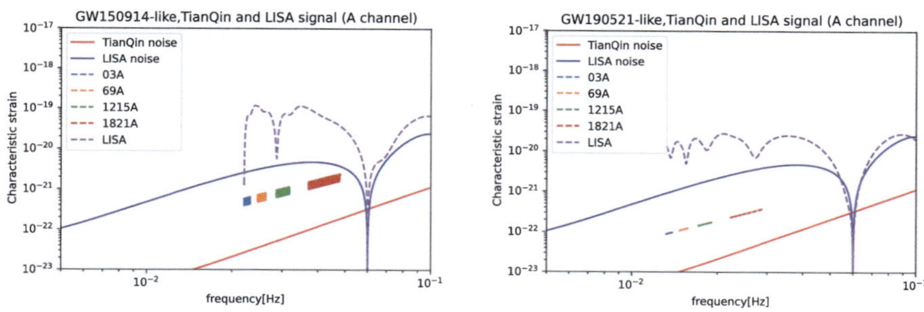

Fig. 9.1 The observation data and noise counterpart of GW150914-like and GW190521-like for TianQin and LISA are displayed in the upper and lower subplot. The purple dashed line and the other four parts dashed lines represent the characteristic strain $2f|\tilde{h}(f)|$ observed by LISA and TianQin in 3+3 mode. The blue and red solid lines are the noise of the LISA and TianQin $\sqrt{fS_n(f)}$

9.3.2 Parameter Estimation Results

For most of the parameters of the sources in Table 9.1, TianQin and LISA have demonstrated similar constraining abilities, which show compatible 1σ widths. This is expected since the sources' SNRs are also comparable.

The above table presents parameter estimation precision for GW190521-like event, the inject waveform values and the priors regions set for each parameter. Space-based GW detectors can constrain black hole spin parameters, but spin effects

Table 9.1 1σ confidence region of parameters and injected parameters' value for the GW190521-like case

Parameters	Injected value	Priors	TianQin	LISA	Joint
$M_c(M_\odot)$	65.10	[60, 70]	$65.10^{+8.087\times10^{-4}}_{-6.077\times10^{-4}}$	$65.10^{+7.43\times10^{-4}}_{-4.53\times10^{-4}}$	$65.10^{+3.47\times10^{-4}}_{-3.48\times10^{-4}}$
η	0.2460	[0.2, 0.25]	$0.228^{+0.014}_{-0.017}$	$0.224^{+0.017}_{-0.016}$	$0.238^{+0.0085}_{-0.0010}$
χ_a	0.71	[−1, 1]	$0.60^{+0.098}_{-0.16}$	$0.61^{+0.090}_{-0.17}$	$0.67^{+0.051}_{-0.089}$
χ_l	−0.02	[−1, 1]	$0.15^{+0.26}_{-0.30}$	$0.17^{+0.28}_{-0.26}$	$0.082^{+0.24}_{-0.22}$
ι(rad)	0.52	[0, π]	$0.48^{+0.40}_{-0.34}$	$0.55^{+0.38}_{-0.38}$	$0.44^{+0.38}_{-0.31}$
D_L(Mpc)	900	[400, 1400]	$972.18^{+178.50}_{-222.44}$	$962.05^{+205.49}_{-221.23}$	$972.18^{+143.45}_{-202.52}$
$t_c - 6.31\times10^7(s)$	0	[-5000, 5000]	-154^{+197}_{-210}	$-59.9^{+53.5}_{-69.7}$	$-22.1^{+51.2}_{-22.8}$
λ(rad)	5.66	[0, 2π]	$5.66^{+0.0061}_{-0.0064}$	$5.66^{+0.0049}_{-0.0049}$	$5.66^{+0.0029}_{-0.0030}$
β(rad)	−0.055	[$-\pi/2, \pi/2$]	$-0.0093^{+0.090}_{-0.074}$	$-0.014^{+0.067}_{-0.058}$	$-0.0046^{+0.062}_{-0.060}$
ψ(rad)	2.094	[0, π]	$1.51^{+1.089}_{-1.015}$	$1.55^{+1.092}_{-1.075}$	$1.67^{+0.97}_{-1.093}$
ϕ_c(rad)	0	[0, 2π]	$0.0067^{+2.057}_{-2.20}$	$0.16^{+2.033}_{-2.17}$	$-0.18^{+2.23}_{-2.15}$

during inspiral are subdominant, leading to strong correlations and imprecise results. For mass parameters (M_c, η), joint observations significantly improve estimates, narrowing 90% confidence intervals. Relative precisions of chirp mass and symmetric mass ratio, $(\Delta M_c)/M_c$ and $\Delta\eta/\eta$, reach $\sim 10^{-7}$ and 0.7 levels, respectively. Long-term signal accumulation by space detectors enables precise phase measurements; as M_c is in the phase term, minor changes significantly impact the phase, allowing good constraints by TianQin and LISA.

For luminosity distance (D_L) and inclination (ι), joint observations do not significantly improve constraints compared to individual detectors due to the strong degeneracy between these parameters in the waveform amplitude. Only their product is well-constrained, and joint observations do not break this degeneracy.

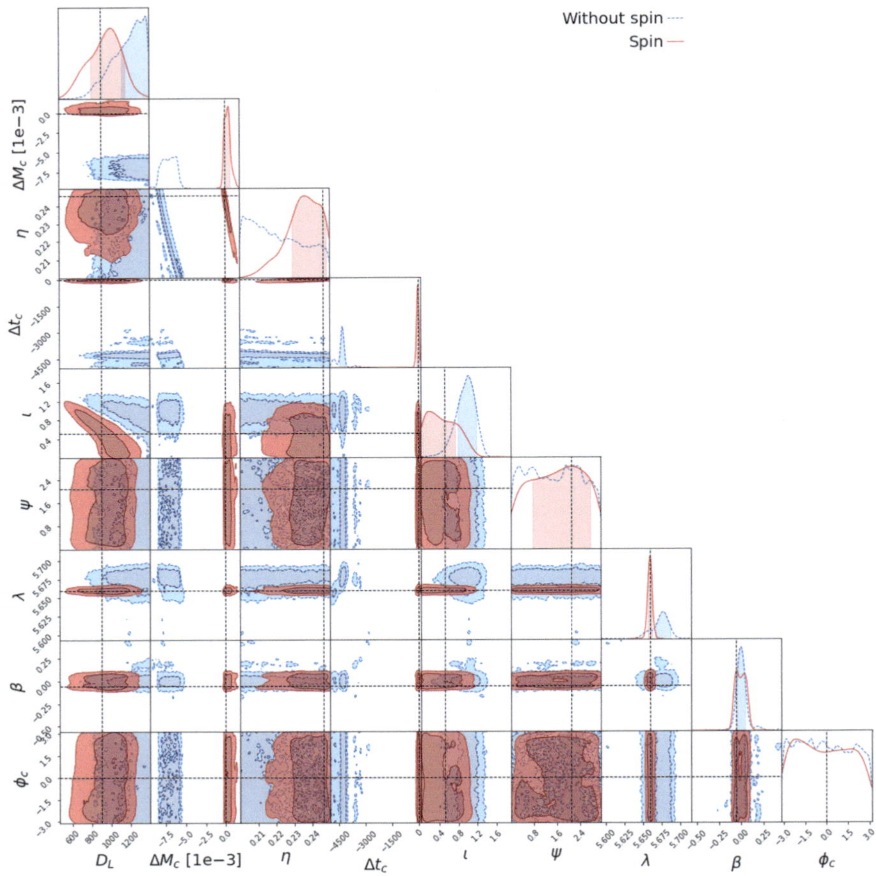

Fig. 9.2 The comparison between GW190521 PE results with considering the spin (spin) in red and PE results with assuming all spin to be 0 (Without spin) in blue. The black dashed lines indicate the true value and the $1-2\sigma$ contour

Joint observations also improve source localization accuracy (sky position angles λ, β) in the Solar System Barycenter(SSB) frame. The precision of coalescence time estimation reaches seconds, demonstrating space detectors' potential for early warnings for electromagnetic follow-ups.

In summary, joint observations enhance parameter constraints compared to individual TianQin/LISA observations, attributed to improved signal-to-noise ratios (Fig. 9.2).

The Fig. 9.2 above shows the parameter estimation results for a GW190521-like signal under joint observations by TianQin and LISA. Specifically, the red contours represent estimates when spin effects are properly accounted for, while the blue contours correspond to incorrect estimates where spin is neglected. The black dashed lines indicate the true values of the gravitational wave signal parameters. It is evident that neglecting spin effects leads to varying degrees of deviation from the true values for different parameters. Notably, the joint distribution of mass parameters (M_c, η) extends beyond the 3σ region of the true values, indicating that mass and spin are interdependent, and omitting spin significantly biases the mass estimates.

References

1. B.P. Abbott. et al., Observation of gravitational waves from a binary black hole merger. Phys. Rev. Lett. **116**(6), 061102 (2016)
2. R. Abbott et al., GWTC-3: compact binary coalescences observed by LIGO and Virgo during the second part of the third observing run. Phys. Rev. X **13**(4), 041039 (2021)
3. P. Amaro-Seoane et al., Laser Interferometer Space Antenna. 2 (2017)
4. J. Luo et al., TianQin: a space-borne gravitational wave detector. Class. Quant. Grav. **33**(3), 035010 (2016)
5. F. Acernese et al., Calibration of advanced Virgo and reconstruction of the gravitational wave signal h(t) during the observing run o2. Class. Quant. Grav. **35**(20), 205004 (2018)
6. A. Sesana, Prospects for multiband gravitational-wave astronomy after GW150914. Phys. Rev. Lett. **116**(23), 231102 (2016)
7. A. Sesana, Multi-band gravitational wave astronomy: science with joint space- and ground-based observations of black hole binaries. J. Phys: Conf. Ser. **840**(1), 012018 (2017)
8. S. Liu, H. Yi-Ming, J. Zhang, J. Mei, Science with the TianQin observatory: preliminary results on stellar-mass binary black holes. Phys. Rev. D **101**(10), 103027 (2020)
9. A. Toubiana, S. Marsat, S. Babak, J. Baker, T. Dal Canton, Parameter estimation of stellar-mass black hole binaries with LISA. Phys. Rev. D **102**, 124037 (2020)
10. A. Nishizawa, A. Sesana, E. Berti, A. Klein, Constraining stellar binary black hole formation scenarios with eLISA eccentricity measurements. Mon. Not. Roy. Astron. Soc. **465**(4), 4375–4380 (2017)
11. R. Buscicchio, A. Klein, E. Roebber, C.J. Moore, D. Gerosa, E. Finch, A. Vecchio, Bayesian parameter estimation of stellar-mass black-hole binaries with LISA. Phys. Rev. D. **104**(4), 044065 (2021)
12. X. Lyu, E.-K. Li, H. Yi-Ming, Parameter estimation of stellar mass binary black holes under the network of TianQin and LISA. Phys. Rev. D. **108**(8), 083023 (2023)
13. C.J. Moore, D. Gerosa, A. Klein, Are stellar-mass black-hole binaries too quiet for LISA?. Mon. Not. Roy. Astron. Soc. **488**(1), 94 (2019)
14. D. Bandopadhyay, C.J. Moore, A GPU-accelerated semi-coherent hierarchical search for stellar-mass binary inspiral signals in LISA (2024). arXiv:2408.13170

15. Y. Fu, Y. Wang, S.D. Mohanty, Hierarchical search method for gravitational waves from stellar-mass binary black holes in noisy space-based detector data (2024), arXiv:2407.10797
16. H. Wang, I. Harry, A. Nitz, Y.-M. Hu, Space-based gravitational wave observatories will be able to use eccentricity to unveil stellar-mass binary black hole formation. Phys. Rev. D **109**(6), 063029 (2024)
17. S. Marsat, J.G. Baker, T. Dal Canton, Exploring the Bayesian parameter estimation of binary black holes with LISA. Phys. Rev. D **103**(8), 083011 (2021)
18. A. Toubiana, S. Marsat, E. Barausse, S. Babak, J. Baker, Tests of general relativity with stellar-mass black hole binaries observed by LISA. Phys. Rev. D **101**(10), 104038 (2020)
19. S. Liu, L.-G. Zhu, H. Yi-Ming, J. Zhang, M.-J. Ji, Capability for detection of GW190521-like binary black holes with TianQin. Phys. Rev. D **105**(2), 023019 (2022)
20. J.W. Armstrong, F.B. Estabrook, M. Tinto. Time-delay interferometry for space-based gravitational wave searches. The Astrophys. J. **527**, 814–826 (1999). ADS Bibcode: 1999ApJ...527..814A
21. S.V. Dhurandhar, K. Rajesh Nayak, J.-Y. Vinet, Algebraic approach to time-delay data analysis for LISA. Phys. Rev. D **65**(10), 102002 (2002)
22. F.B. Estabrook, M. Tinto, J.W. Armstrong, Time-delay analysis of LISA gravitational wave data: elimination of spacecraft motion effects. Phys. Rev. D **62**(4), 042002 (2000)
23. M. Tinto, J.W. Armstrong, Cancellation of laser noise in an unequal-arm interferometer detector of gravitational radiation. Phys. Rev. D **59**(10), 102003 (1999)
24. M. Tinto, F.B. Estabrook, J.W. Armstrong, Time delay interferometry with moving spacecraft arrays. Phys. Rev. D **69**, 082001 (2004)
25. S. Marsat, J.G. Baker, Fourier-domain modulations and delays of gravitational-wave signals. 6 (2018)

Chapter 10
Using Phenomenological Waveforms to Achieve High-Precision Frequency Measurements for Extreme Mass Ratio Inspiral Gravitational Wave Signals

Changqing Ye

Abstract Extreme mass ratio inspirals (EMRIs) are one of the primary gravitational wave sources of astrophysical origin detectable by TianQin. It is crucial to accurately identify EMRI signals with high frequency resolution ($\sim 10^{-6}$ Hz). This paper serves as both an introduction and a supplementary explanation to the phenomenological waveform search discussed in [1]. It highlights the advantages of using phenomenological waveforms with nested sampling methods for high-precision frequency measurement.

10.1 Introduction

On September 14, 2015, the LIGO team detected the gravitational wave signal GW150914 for the first time, thereby ushering in the era of gravitational wave astronomy[2]. Space-based gravitational wave detectors, such as TianQin [3, 4] and LISA[5, 6], are expected to open the millihertz frequency band for gravitational wave observations. This will enable detailed studies of a wide range of gravitational wave signals, providing new insights into fundamental questions in astronomy and physics [3, 5].

In a binary system, a compact object (CO) with a mass ranging from approximately 1–100$M_\odot$ (typically a white dwarf, neutron star, or low-mass black hole), is captured by a massive black hole (MBH) with a mass between 10^4 and $10^7 M_\odot$. The compact object spirals inward towards the MBH, emitting gravitational waves during this process. Due to the extreme mass ratio of $>10^4$, this phenomenon is known as extreme mass ratio inspiral (EMRI). In the inspiral phase, the compact object can orbit the massive black hole around 10^4 to 10^5 cycles [7].

C. Ye (✉)
School of Physics and Electronic Engineering, Sichuan University of Science & Engineering, Zigong, People's Republic of China
e-mail: yechangqing@suse.edu.cn

J. Luo et al. (eds.), *Proceedings of the 7th International Workshop on the TianQin Science Mission; 25–26 April 2024, Hong Kong SAR, China*, Springer Proceedings in Physics 428, https://doi.org/10.1007/978-981-96-7803-7_10

Extreme mass ratio inspirals (EMRIs) are one of the primary gravitational wave sources of astrophysical origin detectable by TianQin[3, 8]. Under optimistic astrophysical models, TianQin is expected to detect tens to hundreds of EMRI events annually[9]. Additionally, TianQin is anticipated to estimate the intrinsic parameters of EMRIs with a precision of 10^{-5} [9, 10]. The high-precision measurements of EMRIs are of significant scientific value.

Achieving high-precision detection targets requires detailed analysis of EMRI signals, which presents significant challenges in data processing. First, the high parameter precision necessitates accurate modeling of EMRI waveforms, as even minor parameter variations (relative errors of 10^{-5}) can lead to completely different waveforms [11, 12]. Hypothetical coverage with template bank requires 10^{40} templates [13]. Second, the multi-peak structure of the EMRI posterior distribution makes it difficult for search algorithms to effectively jump between peaks, complicating the data processing needed to capture the signals and achieve high-precision measurements [14].

The data processing of EMRIs has undergone nearly 20 years of development, resulting in significant advancements. Generally, there are two main approaches. The first is based on template matching, although slower in computation, offers high effectiveness and reliability. Examples of such methods include semi-coherent searches [13], harmonic searches [15], searches using phenomenological waveforms [16], one-step likelihood functions [17], and Markov Chain Monte Carlo (MCMC) methods [14]. The second approach does not rely on template matching, offering faster computation but with certain limitations, such as time-frequency searches [18, 19] and machine learning techniques[20].

Using phenomenological waveforms for signal template matching offers two significant advantages over traditional physical waveforms [16]. First, the generation of physical waveforms requires precise solutions to the small body's orbit and the Teukolsky equation, leading to longer generation times. Second, the degeneracy among the physical intrinsic parameters results in similar phenomenological parameters, creating a many-to-one mapping issue. Therefore, employing phenomenological waveforms in EMRI signal data processing not only significantly reduces computation time but also enables high-precision measurement of the signal's orbital frequencies.

Although previous works, such as time-frequency searches, [18, 19], harmonic searches [15] and phenomenological waveform searches [20], have tracked and identified the frequency evolution of EMRI signals, they offer a relatively low frequency resolution of approximately $\sim 10^{-4}$ Hz. This paper serves as both an introduction and a supplementary explanation to the phenomenological waveform search discussed in [1]. It highlights the advantages of using phenomenological waveforms in conjunction with nested sampling methods for high-precision frequency measurement, where the resolution can reach 10^{-6} Hz. Additionally, some specific details are provided for further clarification.

10.2 Simulating Data

The simulation data consist of the superimposition of EMRI signals and detector noises. The signal is generated by the EMRI waveform via the detection response. The detector noise can be produced as Gaussian noise in line with the noise power spectral density of TianQin.

We adopt the FEW to generate the fully relativistic Schwarzschild eccentricity EMRI waveforms. In this condition, FEW waveforms can quickly generate precise harmonic EMRI waveforms. The waveform $h(t)$ can then be written as [21]

$$h = \frac{\mu}{d_L} \sum_{lmn} A_{lmn}(t)_{-2}Y_{lm}(\theta, \phi) e^{-i\Phi_{mn}(t)}, \tag{10.1}$$

where μ is the mass of the stellar-mass compact object, d_L is the luminosity distance of the source in the observer frame, ${}_{-2}Y_{lm}(\theta, \phi)$ are the regular -2 spin-weighted spherical harmonics functions. The phase $\Phi_{mn} = m\Phi_\varphi + n\Phi_r$ represents the summation of decomposed phases for each given mode,$A_{l,m,n}$ is the amplitude of the corresponding harmonic mode.

In the low-frequency regime ($f < f_*$), the strain $h(t)$ of the detected gravitational wave (GW) signal can be characterized as a linear superposition of the two GW polarizations, $h_+(t)$ and $h_\times(t)$, modulated by the response of the detector[1, 22]:

$$h(t) = F_+ (\theta_S, \phi_S, \psi)\, h_+(t) + F_\times (\theta_S, \phi_S, \psi)\, h_\times(t). \tag{10.2}$$

Here θ_S and ϕ_S are the sky location of the source and $F_{+,\times}$are the so-called antenna pattern functions.

The noise characteristics of the future TianQin detector exhibit a Gaussian distribution and are stationary. The noise is a function of frequency, and the power spectral density(PSD) of the noise can be expressed as [4, 23, 24]:

$$\begin{aligned} S_n(f) =& \frac{1}{L_{\text{arm}}^2} \left[\frac{4S_a}{(2\pi f)^4} \left(\frac{1 + 10^{-4}\ \text{Hz}}{f} \right) + S_x \right] \\ & \left[1 + 0.6 \left(\frac{f}{f_*} \right)^2 \right], \end{aligned} \tag{10.3}$$

where f is the frequency, $S_a = 1 \times 10^{-30}\ \text{m}^2\ \text{s}^{-4}/\text{Hz}$ is the acceleration noise, $S_x = 1 \times 10^{-24}\ \text{m}^2/\text{Hz}$ is the positional noise, and $f_* = \frac{c}{2\pi L_{\text{arm}}}$ is the transfer frequency with c being the speed of light and $L_{\text{arm}} = \sqrt{3} \times 10^8$ m is the arm length.

We consider an EMRI signal with an observation time of 0.5 years, the plunge occurring approximately 0.44 years, and an overall signal-to-noise ratio (SNR) of 50. The specific parameters of this source are detailed in [1]. The distribution of the

Table 10.1 The SNR of the five dominant harmonic modes for the entire observation time and the time segments considered

(m,n)	(2, −2)	(2, −1)	(2, 0)	(2, 1)	(2, 2)	Sum of all modes
[0.0, 0.5] yr	3.59	9.94	21.74	29.07	18.04	50.7
[0.0, 0.1] yr	0.95	1.79	6.97	11.08	7.60	17.01
[0.1, 0.2] yr	0.94	2.50	8.83	9.70	5.56	19.22
[0.2, 0.3] yr	1.40	5.37	11.08	11.51	6.25	22.58
[0.3, 0.4] yr	2.49	5.06	12.70	16.83	10.24	28.78
[0.4, 0.5] yr	1.72	5.91	7.87	14.58	9.64	24.05

SNR for the dominant harmonic modes ($m = 2, n \in [-2, 2]$) of the source, both within each time segment and across the entire observation period, is presented in Table 10.1.

10.3 Search Method

Compared with the physical waveforms, the phenomenological waveforms do not need to solve the gravitational wave waveforms through the orbit of the small celestial body and the Teukolsky equation. Therefore, the calculation speed has been greatly improved. For the EMRI system, in the early stage of the inspiral phase, the evolution of the orbital frequency is relatively slow. The phase $\Phi_{mn}(t)$ can be Taylor expanded as [16]

$$\begin{aligned}\Phi_{mn}(t) =&\Phi_{mn}(t_0) + \omega_{m,n}(t_0)(t - t_0) + \frac{1}{2}\dot{\omega}_{m,n}(t_0) \\ &\times (t - t_0)^2 + \frac{1}{3}\ddot{\omega}_{m,n}(t_0)(t - t_0)^3 + O(4),\end{aligned} \tag{10.4}$$

where the angular frequency $\omega_{m,n} =: m\omega_\phi + n\omega_r$, and $\dot{\omega}_{m,n}$ and $\ddot{\omega}_{m,n}$ are its first and second time-derivatives, respectively.

In the process of searching for signals using phenomenological waveforms, we employed a semi-coherent search method. The semi-coherent search method, which divides data into short time segments, is widely applied in gravitational wave data processing, such as in LIGO and Virgo, as it can significantly reduce the number of parameter templates. However, as the observation time increases, the phase approximation of the phenomenological waveform gradually deviates. Therefore, when performing searches with phenomenological waveforms, only shorter time segments can be analyzed. Nonetheless, if the time segments are too short, the signal-to-noise ratio may be insufficient to distinguish between signal and noise. Consequently, we opted for time segments of approximately 0.1 years for our analysis.

We sample the posterior distributions of the phenomenological parameters for each harmonic mode within each time segment. The standard Bayesian (log-)likelihood ratio function of the source parameters θ, is given by

$$\Lambda(\theta) := (d \mid h) - \frac{1}{2}(h \mid h) \tag{10.5}$$

To perform this sampling, we employ the `dynesty` sampler. Dynamic nested sampling represents an advancement of nested sampling methods by adaptively allocating samples to enhance the efficiency of sampling the target function within the parameter space. The core advantages of this sampler is to continuously update the sample distribution during the process to explore the parameter space more effectively, while minimizing sampling in low-probability regions, thereby improving the overall efficiency of the algorithm.

10.4 Results and Discussion

10.4.1 Results

In [1], we present the posterior distributions of the phenomenological parameters for all three harmonic modes across the three time segments. The posterior distribution of the phenomenological parameters for the (2, 0) mode during the first time segment is shown graphically. Similarly, in Fig. 10.1, we show the posterior distribution of the phenomenological parameters for the (2, 1) mode in the first time segment. It can be seen that, in this scenario, the measurement precision of the initial angular frequency ω_0 can reach the level of $\sim 10^{-4}$, corresponding to a precision of $\sim 10^{-5}$ for the initial frequency f_0. while the precision for Initial angular frequency derivative $\dot{\omega}_0$ and second-order derivative $\ddot{\omega}_0$ can reach the level of 10^{-2} and 10^{-1}, corresponding to a precision of $\sim 10^{-3}$ and $\sim 10^{-2}$ for the initial frequency derivative $\dot{f}_0$ and second-order derivative $\ddot{f}_0$, respectively. It can be inferred that the search employing phenomenological waveforms can achieve an angular frequency ω resolution of $\sim 10^{-6}$ rad/s, corresponding to the frequency f resolution of $\sim 10^{-7}$ Hz.

Based on the measurement results of the harmonic modes from other time segments in [1], the angular frequency ω resolution can reach a level from $\sim 10^{-4}$ rad/s to $\sim 10^{-6}$ rad/s, corresponding to a frequency f resolution from $\sim 10^{-5}$ Hz to $\sim 10^{-7}$ Hz. Therefore, it can be said that we achieved high-precision measurement of the EMRI signal frequency by using the phenomenological waveform.

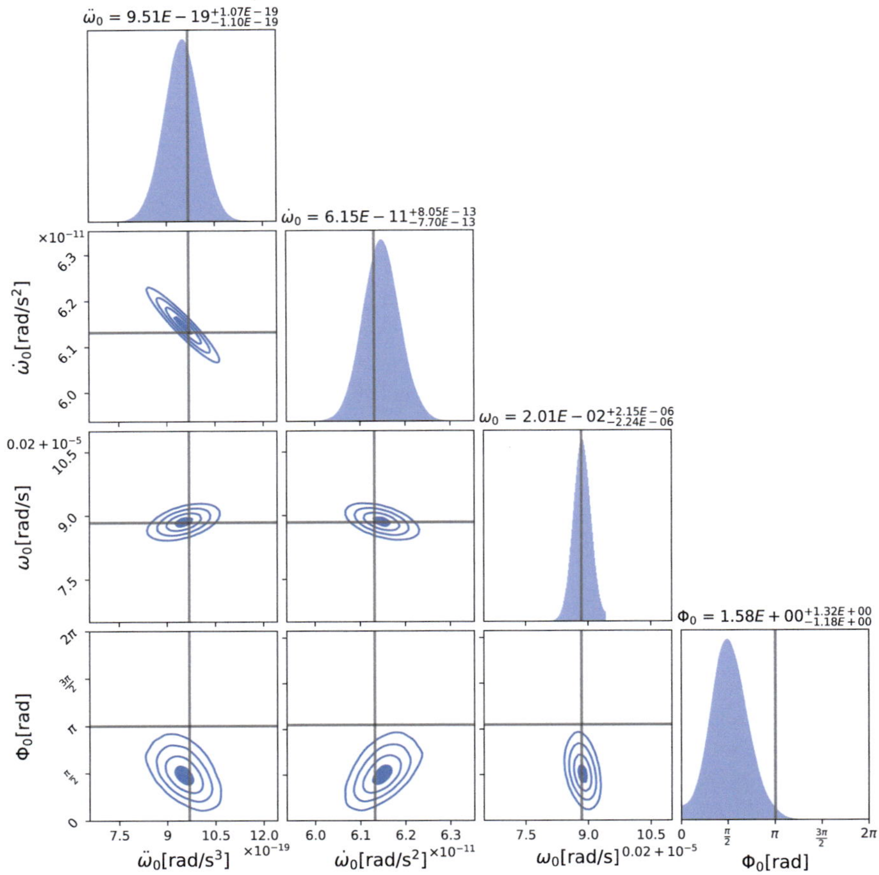

Fig. 10.1 The posterior distribution of the four phenomenological parameters for the first time segment (2, 1) mode

10.4.2 Discussion

As noted in [1], the measurement results for certain phenomenological parameters were not as accurate as expected. For example, in the third time segment, the measurement precision of the parameter ω_0 was about 10%. This discrepancy likely arises because, as the signal approaches the plunge, the phenomenological waveform deviates more from the true waveform, leading to measurement errors. Additionally, in the (2, 2) mode of the second time segment, the measurement precision of the parameter Omega was about 2%, which is an order of magnitude lower than that of other modes in the same time segment. We speculate that this is due to the use of the Taylor approximation method in fitting the phase, where not all phases were accurately fitted, causing the fitting error to exceed the measurement error. A better polynomial fitting method should be chosen.

In the future, phenomenological waveforms, due to their rapid generation capabilities, are expected to be widely applied in waveform modeling and signal detection during the early inspiral phases of EMRIs.

References

1. C.-Q. Ye, H.-M. Fan, A. Torres-Orjuela, J.-D. Zhang, Y.-M. Hu, Identification of gravitational waves from extreme-mass-ratio inspirals. Phys. Rev. D **109**(12), 124034 (2024)
2. B.P. Abbott, R. Abbott, T. Abbott, M. Abernathy, F. Acernese, K. Ackley, C. Adams, T. Adams, P. Addesso, R. Adhikari et al., Gw150914: the advanced ligo detectors in the era of first discoveries. Phys. Rev. Lett. **116**(13), 131103 (2016)
3. J. Luo et al., TianQin: a space-borne gravitational wave detector. Class. Quant. Grav. **33**(3), 035010 (2016)
4. J. Mei, Y.-Z. Bai, J. Bao, E. Barausse, L. Cai, E. Canuto, B. Cao, W.-M. Chen, Y. Chen, Y.-W. Ding et al., The tianqin project: current progress on science and technology. Prog. Theor. Exp. Phys. (2020)
5. K. Danzmann, A. Rüdiger, Lisa technology-concept, status, prospects. Class. Quantum Gravity **20**(10), S1 (2003)
6. P. Amaro-Seoane et al., Laser Interferometer Space Antenna, 2 (2017)
7. P. Amaro-Seoane, Relativistic dynamics and extreme mass ratio inspirals. Living Rev. Rel. **21**(1), 4 (2018)
8. A. Torres-Orjuela, S.-J. Huang, Z.-C. Liang, S. Liu, H.-T. Wang, C.-Q. Ye, Y.-M. Hu, J. Mei, Detection of astrophysical gravitational wave sources by TianQin and LISA (2023). arXiv:2307.16628
9. H.-M. Fan, Y.-M. Hu, E. Barausse, A. Sesana, J.-D. Zhang, X. Zhang, T.-G. Zi, J. Mei et al., Science with the tianqin observatory: preliminary result on extreme-mass-ratio inspirals. Phys. Rev. D **102**(6), 063016 (2020)
10. T.-G. Zi, J.-D. Zhang, H.-M. Fan, X.-T. Zhang, Y.-M. Hu, C. Shi, J. Mei, Science with the TianQin observatory: preliminary results on testing the no-hair theorem with extreme mass ratio inspirals. Phys. Rev. D **104**(6), 064008 (2021)
11. L. Barack, C. Cutler, LISA capture sources: Approximate waveforms, signal-to-noise ratios, and parameter estimation accuracy. Phys. Rev. D **69**(8), 082005 (2004)
12. A.J. Chua, M.L. Katz, N. Warburton, S.A. Hughes, Rapid generation of fully relativistic extreme-mass-ratio-inspiral waveform templates for lisa data analysis. Phys. Rev. Lett. **126**(5), 051102 (2021)
13. J.R. Gair, L. Barack, T. Creighton, C. Cutler, S.L. Larson, E.S. Phinney, M. Vallisneri, Event rate estimates for lisa extreme mass ratio capture sources. Class. Quantum Gravity **21**(20), S1595 (2004)
14. A.J. Chua, C.J. Cutler, Non-local parameter degeneracy in the intrinsic space of gravitational-wave signals from extreme-mass-ratio inspirals (2021). arXiv:2109.14254
15. S. Babak, J.R. Gair, E.K. Porter, An algorithm for the detection of extreme mass ratio inspirals in lisa data. Class. Quantum Gravity **26**(13), 135004 (2009)
16. Y. Wang, Y. Shang, S. Babak, Extreme mass ratio inspiral data analysis with a phenomenological waveform. Phys. Rev. D **86**(10), 104050 (2012)
17. A.J. Chua, One-stop function for gravitational-wave detection, identification, and inference. Phys. Rev. D **106**(10), 104051 (2022)
18. L. Wen, J.R. Gair, Detecting extreme mass ratio inspirals with lisa using time-frequency methods. Class. Quantum Gravity **22**(10), S445 (2005)
19. J. Gair, L. Wen, Detecting extreme mass ratio inspirals with lisa using time–frequency methods: Ii. search characterization. Class. Quantum Gravity **22**(18), S1359 (2005)

20. X.-T. Zhang, C. Messenger, N. Korsakova, M.L. Chan, Y.-M. Hu, J.-D. Zhang, Detecting gravitational waves from extreme mass ratio inspirals using convolutional neural networks. Phys. Rev. D **105**(12), 123027 (2022)
21. M.L. Katz, A.J. Chua, L. Speri, N. Warburton, S.A. Hughes, Fast extreme-mass-ratio-inspiral waveforms: New tools for millihertz gravitational-wave data analysis. Phys. Rev. D **104**(6), 064047 (2021)
22. X.-C. Hu, X.-H. Li, Y. Wang, W.-F. Feng, M.-Y. Zhou, Y.-M. Hu, S.-C. Hu, J.-W. Mei, C.-G. Shao, Fundamentals of the orbit and response for TianQin. Class. Quant. Grav. **35**(9), 095008 (2018)
23. J. Luo, L.-S. Chen, H.-Z. Duan, Y.-G. Gong, S. Hu, J. Ji, Q. Liu, J. Mei, V. Milyukov, M. Sazhin et al., Tianqin: a space-borne gravitational wave detector. Class. Quantum Gravity **33**(3), 035010 (2016)
24. A. Torres-Orjuela, S.-J. Huang, Z.-C. Liang, S. Liu, H.-T. Wang, C.-Q. Ye, Y.-M. Hu, J. Mei, Detection of astrophysical gravitational wave sources by tianqin and lisa. Sci. China Phys. Mech. & Astron. **67**(5), 259511 (2024)

Chapter 11
Time Delay Interferometry in Space-Based Gravitational Wave Detection

Pan-Pan Wang and Cheng-Gang Shao

Abstract The space-based gravitational wave detector will open a new frequency window ranging from 0.1mHz−1Hz, which is difficult to access from the ground-based detectors, promising to provide crucial information for studying fundamental physics related to astrophysics and early universe processes. However, gravitational wave signals are extremely weak, and any noise during the detection process can potentially overwhelm these signals. Therefore, suppressing noise is key to the successful detection of gravitational waves. Currently, laser frequency noise and clock jitter noise are approximately seven and three orders of magnitude larger in intensity than that of typical gravitational waves, respectively. These two major noise sources cannot be directly overcome by experimental techniques within the short term due to the physical limitations of the instruments. Therefore, it is necessary to employ the Time Delay Interferometry (TDI) technique in data pre-processing, which involves delaying and recombining data stream, to suppress these noises. This paper will introduce methods for obtaining various TDI combinations, including algebraic methods for solving indeterminate equations and geometric methods for symbolic search. Based on the multiple TDI types, we will analyze the technique of constructing independent measurement data to reduce clock noise using clock sideband comparison. Finally, we will present the sensitivity functions for different types of TDI combinations, which are used to assess the sensitivity limits of the TDI combinations.

P.-P. Wang (✉) · C.-G. Shao
National Gravitation Laboratory, MOE Key Laboratory of Fundamental Physical Quantities Measurement, and School of Physics, Huazhong University of Science and Technology, Wuhan 430074, People's Republic of China
e-mail: ppwang@hust.edu.cn

C.-G. Shao
e-mail: cgshao@hust.edu.cn

J. Luo et al. (eds.), *Proceedings of the 7th International Workshop on the TianQin Science Mission; 25–26 April 2024, Hong Kong SAR, China*, Springer Proceedings in Physics 428, https://doi.org/10.1007/978-981-96-7803-7_11

11.1 Laser Interferometry Measurement Data and Conventions

For a typical space-based gravitational wave detector, three spacecraft form an equilateral triangle. Each spacecraft consists of two movable optical subassemblies (MOSAs), two lasers, one clock and other components, as shown in Fig. 11.1. Gravitational wave detection employs a three-stage measurement principle, namely the test mass to optical bench, bench to bench, and bench to test mass, achieved through science interferometer, test mass interferometer, and reference interferometer, respectively [1].

To facilitate the representation of the time-varying delay process between spacecraft, we first define six delay operators D_i, $D_{i'}$ $(i = 1, 2, 3)$. These operators shift all time-dependent functions by the corresponding light travel times L_i, $L_{i'}$ respectively. In this paper, we adopt the convention $c = 1$, for any data stream $f(t)$, we have:

$$
\begin{aligned}
D_i f(t) &= f(t - L_i(t)),\\
D_j D_i f(t) &= D_{ji} f(t) = f(t - L_i(t - L_j) - L_j(t)).
\end{aligned}
\tag{11.1}
$$

According to the principle of laser exchanges between the three spacecraft, each spacecraft has four different laser interferometry measurement data streams, which are: (1) Scientific carrier interferometric data stream: the interferometry of incident laser from a distant spacecraft and local laser. The incident laser carries gravitational wave information, and this interferometry measurement signal is denoted as $s_i^c(t)$. (2) Test mass interferometric data stream: the laser exchange between two adjacent

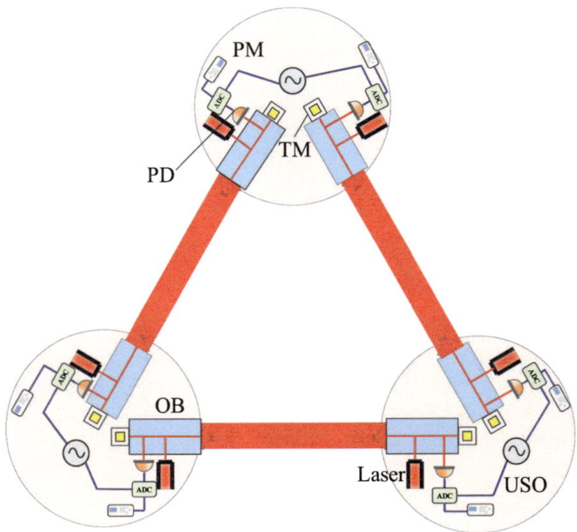

Fig. 11.1 The constellation configuration of space-borne gravitational wave detector. OB: optical bench; USO: ultra-stable oscillator; PD: photodetector; TM: test mass

optical benches on the same spacecraft. The laser from optical bench i' is transmitted via optical fiber to the test mass of optical bench i, where it is reflected and interfered with the local laser. This data stream is denoted as $\epsilon_i(t)$. (3) Reference interferometric data stream: the laser from the adjacent optical bench i' is transmitted via optical fiber to optical bench i and heterodyned with the local laser beam, resulting in the reference interferometry data stream $\tau_i(t)$. (4) Sideband data stream: to eliminate clock noise, electro-optic modulators generate sidebands at the ends of the carrier. The driving frequency of the electro-optic modulator is referenced to the clock, transferring the clock noise to the remote spacecraft. The sidebands are heterodyned to produce the sideband data stream $s_i^{sb}(t)$.

For simplicity, only the following noises are considered for each data stream on the spacecraft: laser frequency noise, clock noise, optical bench noise, test mass acceleration noise, fiber noise, and optical measurement system noise.

On the optical bench i, these are denoted as follows [2]:

$$\begin{aligned} s_i^c =& h_i + D_{i-1} p_{(i+1)'} - p_i + N_i^{oms} - a_i q_i + \\ & 2\pi \nu_{(i+1)'} \left[\vec{n}_{i-1} \cdot D_{i-1} \vec{\Delta}_{(i+1)'} + \vec{n}_{(i-1)'} \cdot \vec{\Delta}_i \right], \\ s_i^{sb} =& h_i + D_{i-1} p_{(i+1)'} - p_i + N_i^{oms,sb} - c_i q_i - m_i q_i + \\ & 2\pi \nu_{(i+1)'} \left[\vec{n}_{i-1} \cdot D_{i-1} \vec{\Delta}_{(i+1)'} + \vec{n}_{(i-1)'} \cdot \vec{\Delta}_i \right] + m_{(i+1)'} D_{i-1} q_{i+1}, \\ \varepsilon_i =& p_{i'} - p_i + \mu_{i'} - 4\pi \nu_{i'} \left[\vec{n}_{(i-1)'} \cdot \vec{\delta}_i - \vec{n}_{(i-1)'} \cdot \vec{\Delta}_i \right] - b_i q_i, \\ \tau_i =& p_{i'} - p_i + \mu_{i'} - b_i q_i. \end{aligned} \tag{11.2}$$

On the optical bench i', these are represented as:

$$\begin{aligned} s_{i'}^c =& h_{i'} + D_{(i+1)'} p_{i-1} - p_{i'} + N_{i'}^{oms} - a_{i'} q_{i'} + \\ & 2\pi \nu_{i-1} \left[\vec{n}_{i+1} \cdot D_{(i+1)'} \vec{\Delta}_{i-1} + \vec{n}_{i+1} \cdot \vec{\Delta}_{i'} \right], \\ s_{i'}^{sb} =& h_{i'} + D_{(i+1)'} p_{i-1} - p_{i'} + N_{i'}^{oms,sb} - c_{i'} q_i - m_{i'} q_i + \\ & 2\pi \nu_{i-1} \left[\vec{n}_{(i+1)'} \cdot D_{(i+1)'} \vec{\Delta}_{i-1} + \vec{n}_{i+1} \cdot \vec{\Delta}_{i'} \right] + m_{i-1} D_{(i+1)'} q_{i-1}, \\ \varepsilon_{i'} =& p_i - p_{i'} + \mu_i - 4\pi \nu_{i'} \left[\vec{n}_{i+1} \cdot \vec{\delta}_{i'} - \vec{n}_{i+1} \cdot \vec{\Delta}_{i'} \right] - b_{i'} q_i, \\ \tau_{i'} =& p_i - p_{i'} + \mu_i - b_{i'} q_i, \end{aligned} \tag{11.3}$$

where h_i represents the gravitational wave signal, p_i denotes the laser frequency noise, ν_i is the center frequency of the laser, $\vec{\Delta}_i$ refers to the optical bench noise, N_i^{oms} represents the optical measurement system noise, q_i stands for the clock noise, μ_i denotes the fiber noise, $\vec{\delta}_i$ represents the test mass acceleration noise, and m_i is the modulation factor. The subscripts follow a cyclic pattern $1 \rightarrow 2 \rightarrow 3 \rightarrow 1$.

The conversion coefficients $a_{i,i'}$, $b_{i,i'}$, $c_{i,i'}$ are defined as:

$$
\begin{aligned}
a_i &= \frac{\nu_{(i+1)'}\left(1-\dot{L}_{i-1}\right)-\nu_i}{f_i}, a_{i'} = \frac{\nu_{i-1}\left(1-\dot{L}_{(i+1)'}\right)-\nu_{i'}}{f_i}, \\
b_i &= \frac{\nu_{i'}-\nu_i}{f_i}, b_{i'} = \frac{\nu_i-\nu_{i'}}{f_i} = -b_i, \\
c_i &= \frac{\left(\nu_{(i+1)'}+m_{(i+1)'}f_{i+1}\right)\left(1-\dot{L}_{i-1}\right)-\left(\nu_i+m_i f_i\right)}{f_i}, \\
c_{i'} &= \frac{\left(\nu_{i-1}+m_{i-1}f_{i-1}\right)\left(1-\dot{L}_{(i+1)'}\right)-\left(\nu_{i'}+m_{i'}f_i\right)}{f_i},
\end{aligned}
\tag{11.4}
$$

where f_i is clock frequency, $\dot{L}$ is the first-order time derivatives of arm length.

The interferometry data stream, after being digitized by clock-driven analog-digital converters (ADCs), are fed into the phase measurement system. The recorded data streams are then transmitted to Earth, where the received data streams are processed to extract gravitational wave signals.

We introduce intermediate variables to eliminate optical bench noise and the laser frequency noise with primed. Ultimately, we obtain [2, 3]:

$$
\begin{aligned}
\eta_i =& h_i + D_{i-1}p_{i+1} - p_i - a_i q_i + D_{i-1}b_{i+1}q_{i+1} + N_i^{oms} \\
&+ 2\pi\,\nu_{(i+1)'}\vec{n}_{i-1}\left[D_{i-1}\vec{\delta}_{(i+1)'} - \vec{\delta}_i\right], \\
\eta_{i'} =& h_{i'} + D_{(i+1)'}p_{i-1} - p_i + \left(b_{i'} - a_{i'}\right)q_i + N_{i'}^{oms} \\
&+ 2\pi\,\nu_{i-1}\vec{n}_{i+1}\cdot\left[\vec{\delta}_i - D_{(i+1)'}\vec{\delta}_{i-1}\right].
\end{aligned}
\tag{11.5}
$$

The above expression assumes that the fiber noise is reciprocal.

The main noise in the data streams can be divided into two categories, one is the inevitable noise source, which we regard as the noise floor. The other is technical noise. Some noises can be avoided by high-precision experimental technology, but laser noise and clock noise are several orders of magnitude higher than gravitational wave signal, it is impossible to avoid the influence of these two noises by relying on current experimental technology. Therefore, data pre-processing is needed to reduce the influence of two noises on gravitational wave detection. Next, TDI to suppress these two noises will be introduced.

11.2 Time Delay Interferometry

For an equal-arm interferometer, since the laser traverses two arms of equal length, we can measure gravitational wave strains of 10^{-20} with laser stable at 10^{-13}. However, if the arm lengths are unequal, residual laser noise will be present.

TDI technique constructs an effective equal-arm interferometer through an appropriate linear combination of time-shifted data streams [4–9].

From the perspective of mathematical expressions, for a Michelson interferometer with unequal arms, as shown in Fig. 11.2(a), after the laser is split by the beam splitter, it interferes with the local laser after traveling back and forth along two unequal arm lengths to obtain two measurement data, $y_{PD,1}(t)$ and $y_{PD,2}(t)$. Due to the unequal arm lengths, there is residual laser noise after subtracting the two measurement data. If the two measurement data are separately considered delayed $2L_2$, $2L_1$ and then differenced, there will still be residual laser noise. If the differential signal of the measurement data $y_{PD,1}(t) - y_{PD,2}(t)$ is combined with the differential signal of the delayed data $y_{PD,1}(t - 2L_2) - y_{PD,2}(t - 2L_1)$, then the combination precisely cancels out the laser frequency noise, as shown in Fig. 11.2(b).

Replacing the mirrors of the Michelson interferometer with spacecrafts gives rise to the situation of space-based gravitational wave detection. Unlike the two measurement data streams, $y_{PD,1}(t)$ and $y_{PD,2}(t)$, of the Michelson interferometer, data streams will also be obtained at the two spacecraft. Therefore, the Michelson TDI combination utilizes the delayed combination of four data streams η_1, $\eta_{2'}$, $\eta_{1'}$, and η_3, to eliminate laser frequency noise.

In space-based gravitational wave detection, the general expression for TDI is [1, 10–16]:

$$\text{TDI} = \sum_{i=1}^{3}(P_i\eta_i + P_{i'}\eta_{i'}), \tag{11.6}$$

where $P_{i,i'}$ are the polynomial coefficients of the time delay operator.

After utilizing TDI, the clock noise is two orders of magnitude larger than the noise floor, as shown in Fig. 11.3, it is necessary to develop clock noise suppression algorithm. To suppress clock noise, additional data streams need to be introduced, namely the sideband data stream. By observing the differences between the sideband data and the carrier data, it is found that they mainly differ in terms of clock noise, one finds [2, 17]

$$r_i = q_i - D_{i-1}q_{i+1}, r_{i'} = q_i - D_{(i+1)'}q_{i-1}. \tag{11.7}$$

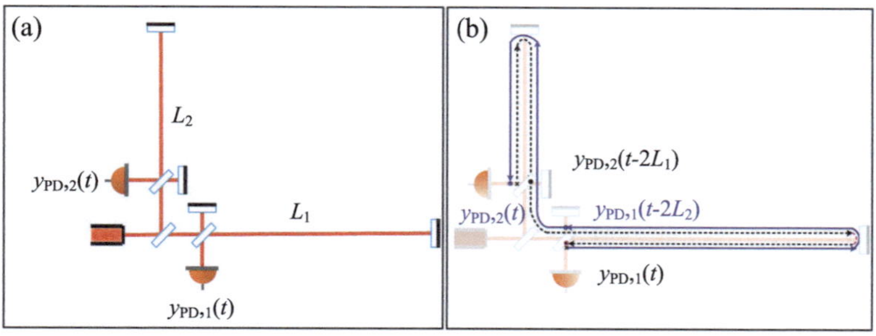

Fig. 11.2 **a** Schematic diagram of Michelson laser interferometer, **b** Schematic diagram of virtual equal arm interferometer

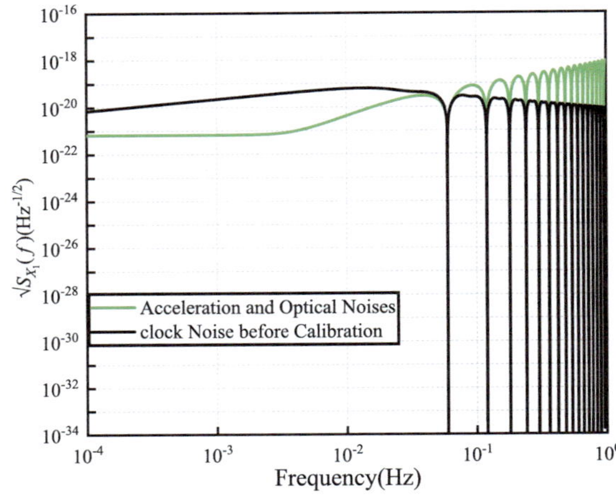

Fig. 11.3 The noise floor and residual clock noise curves of the first-generation Michelson combination, where green curve represents the noise floor and black curve represents the residual clock noise after TDI

Subtracting sideband data from carrier data constructs an independent measurement quantity solely containing clock noise. Further, by time-shifting the $r_{i,i'}$ data and linearly combining them, a clock noise term is constructed, which is then subtracted to remove the clock noise.

To evaluate the performance of different TDI combinations, it is necessary to study the sensitivity function of TDI, which is defined as the ratio of the noise power spectrum to the signal response function.

11.3 TDI and Its Applications

Our main research questions are these three: how to obtain TDI combinations, how to suppress clock noise, and how to assess noise reduction capability. We will now proceed to introduce each of these issues separately.

11.3.1 Algebraic Method and Geometric Method

Firstly, concerning algebraic method for obtaining TDI combinations. The form of TDI is expressed as a polynomial of time delay operators multiplied by data streams. The purpose of TDI combination is to reduce the influence of laser frequency noise. Hence, we can derive the TDI equation [4, 5, 18–20]:

$$\begin{aligned}
P_1 + P_{1'} - P_{2'} D_{3'} - P_3 D_2 &= 0, \\
P_2 + P_{2'} - P_{3'} D_{1'} - P_1 D_3 &= 0, \\
P_3 + P_{3'} - P_{1'} D_{2'} - P_2 D_1 &= 0.
\end{aligned} \tag{11.8}$$

Obviously, the solutions of P_i and $P_{i'}$ are polynomials in terms of D_i and $D_{i'}$. Due to the presence of operators in the TDI equation, we cannot directly perform the operation of moving terms and taking inverses. In terms of equation form, introducing constraint equations categorizes specific types of TDI combinations, we have [18, 19]

$$\begin{aligned} &P_1 + P_{1'} - P_{2'}D_{3'} - P_3 D_2 = 0, \\ &P_2 + P_{2'} - P_{3'}D_{1'} - P_1 D_3 = 0, \\ &P_3 + P_{3'} - P_{1'}D_{2'} - P_2 D_1 = 0, \\ &f(P_1, P_2, P_3, P_{1'}, P_{2'}, P_{3'}) = 0, \\ &g(P_1, P_2, P_3, P_{1'}, P_{2'}, P_{3'}) = 0. \end{aligned} \tag{11.9}$$

Originally, we had three equations to solve six unknowns. After introducing two constraint equations, f and g, we have five equations to solve six unknowns. Clearly, the solutions obtained from the constrained equations are solutions of the original equations. Therefore, by solving the constrained TDI equations, we can solve for the second generation TDI combinations. By observing the configurations of different TDI combinations, constraint equations can be derived. Taking the Michelson combination as an example, it lacks two data streams corresponding to one arm, one finds

$$P_i = P_{(i+1)'} = 0, \tag{11.10}$$

resulting in coefficients of zero for two data streams, which serves as an additional constraint equation.

In addition to using algebraic methods to obtain TDI combinations, employing geometric methods to search the spacetime diagram can also effectively yield various types of TDI combinations [6–9]. It's worth noting that earlier geometric methods utilized two-dimensional diagram containing only spacecraft indices [6]. Afterwards, they have evolved into three-dimensional space-time diagrams containing both time and space indices, as shown in Fig. 11.4. This represents a significant improvement as it more accurately reflects the TDI properties, specifically the two virtual equal-arm paths, forming a closed loop.

Using spacetime diagrams to represent the propagation paths of lasers between spacecraft, it's evident that they possess two degrees of freedom: one for the laser propagation direction along the spacecraft and the other for the time direction, from the spacetime diagram, it can be observed that when the previous link is determined, there are three choices for the next link. This is because if we return to the original situation, the effects are cancelled out. Therefore, we adopt a ternary convention for searching. From another perspective, ternary search also reduces search time. The principle of the search is that TDI combinations must form a closed loop, so that noise is eliminated one by one in a relay-like manner.

Using the ternary symbol search algorithm, we have identified 45 core second-generation TDI combination [21].

In addition to discovering some conventional combinations, we have also identified multiple novel TDI combinations, such as the zero-response combination [22].

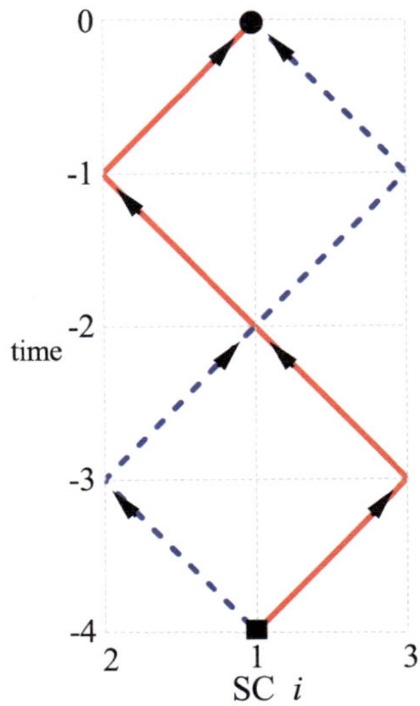

Fig. 11.4 The spacetime diagram of the first-generation Michelson combination.The black squares and dots indicate the initial and terminal grids of the two routes, and the reference time $t = 0$ is attributed to the terminal SC1. The red-solid and blue dashed line segments correspond to the two synthesized routes

In these configurations, the laser paths cancel each other out in the same link, resulting in both instrument noise and response functions being zero. This is highly useful for identifying instrument noise.

11.3.2 Clock Noise Suppression Algorithm

To correct clock noise, it is necessary to perform comparison measurements of the main clocks on different spacecraft [23–25]. To achieve this, the method of using laser phase modulation to imprint each spacecraft's clock signal onto the outgoing laser has been proposed. This process creates sidebands on the laser carrier.

In this way, inter-spacecraft laser links facilitate the interaction of clock signals, enabling the comparison measurements between different clocks and providing the unique measurements needed to correct clock noise. The variable $r_{i,i'}(i = 1, 2, 3)$ represents the clock noise for each of the six one-way interferometric measurements [26]. Each measurement includes local clock noise and remote clock noise delayed by the arm length. Any form of virtual photon path configuration can be synthesized:

$$q_B - D_{i_n i_{n-1} \cdots i_1} q_A. \tag{11.11}$$

By combining multiple similar expressions, the form of the delay operator polynomial can be constructed as

$$\sum_i P_i q_i. \tag{11.12}$$

From $r_{i,i'}$, we can construct six expressions for $R_{i,i'} (i = 1, 2, 3)$ as follows:

$$R_i = P_i q_i, \ R_{i'} = P_{i'} q_i. \tag{11.13}$$

The general expression for any TDI variable satisfying the above conditions is

$$\begin{aligned} \mathrm{TDI}^q &= -\sum_{i=1}^{3} (a_i P_i + a_{i'} P_{i'} - b_{i'} \left[P_{i'} - P_{i-1} D_{i+1}\right]) q_i \\ &= -\sum_{i=1}^{3} \left[\left(a_i + b_{(i+1)'} \right) R_i + \left(a_{i'} - b_{i'} \right) R_{i'} - b_{(i+1)'} P_i r_i \right]. \end{aligned} \tag{11.14}$$

Thus, the combination for eliminating clock noise is in the form of

$$\begin{aligned} \mathrm{TDI}^c =& \sum_{i=1}^{3} (P_i \eta_i + P_{i'} \eta_{i'}) - \\ & \left[\sum_{i-1}^{3} b_{(i+1)'} P_i r_i - (a_i + b_{(i+1)'}) R_i - (\alpha_{i'} - b_{i'}) R_{i'} \right]. \end{aligned} \tag{11.15}$$

After utilizing this expression to eliminate clock noise, it meets the requirements for space-based gravitational wave detection.

11.3.3 Sensitivity Function of TDI Combinations

In the Sec.3.1, algebraic and geometric methods are used to present more effective multi-type TDI combinations for suppressing laser frequency noise. The Sec.3.2, utilizing sideband data streams, general strategies for suppressing clock noise are given.

To evaluate the noise reduction performance of different TDI combinations, it is essential to understand the sensitivity limits of the instrument [27, 28]. Typically, the sensitivity function depends on the ratio of the instrument's noise power spectral density (PSD) to the signal response, representing the strength of noise translated into the signal. It is defined as [29–32]:

$$S(u) \equiv B \frac{\sqrt{N(u)}}{\sqrt{C_A \sqrt{R(u)}}} \sqrt{\frac{1}{T}}, \tag{11.16}$$

Here, B represents the multiple of the signal-to-noise ratio, $N(u)$ is the PSD of the noise floor, C_A is the average over the orbital inclination, $R(u)$ is the average response function of the instrument to the gravitational wave source, and T denotes the accumulated signal time.

No matter what polarization mode the gravitational wave is in, the instrument noise remains consistent. After using TDI techniques to suppress laser frequency noise, the PSD of the test mass acceleration noise and the optical measurement system noise is written as

$$\begin{aligned} N(u) &= S^{a}_{\mathrm{TDI}}(u) + S^{oms}_{\mathrm{TDI}}(u) \\ &= C_1[\widetilde{P}_i(u)]n_1(u) + 2C_2[\widetilde{P}_i(u)]n_2(u), \end{aligned} \tag{11.17}$$

where

$$\begin{aligned} C_1[\tilde{P}_i(u)] &= \sum_{i=1}^{3}\left[\left|\tilde{P}_i\right|^2 + \left|\tilde{P}_{i'}\right|^2\right], \\ C_2[\tilde{P}_i(u)] &= \sum_{i=1}^{3}\mathrm{Re}\left[\tilde{P}_i\tilde{P}^*_{(i+1)'}\right], \end{aligned} \tag{11.18}$$

and

$$\begin{aligned} n_1(u) &= 2 \times S_a + S_{oms}, \\ n_2(u) &= S_a \cos u, \end{aligned} \tag{11.19}$$

where S_a and S_{oms} are the PSD for test mass acceleration noise and optical measurement system noise, $u = \frac{2\pi f L}{c}$ is a dimensionless quantity. f is the frequency of gravitational waves, L the arm length of the detector, c is the speed of light.

Regarding the response function, space-based gravitational wave detector has six arms. We first provide the response function of a single data stream. The response function for a specific TDI combination is shown as:

$$F_{+,\times} = P_1 F_{\eta_1;+,\times} + P_2 F_{\eta_2;+,\times} + P_3 F_{\eta_3;+,\times} + P_{1'} F_{\eta_{1'};+,\times} + P_{2'} F_{\eta_{2'};+,\times} + P_{3'} F_{\eta_{3'};+,\times}. \tag{11.20}$$

For ease of calculation, we perform a all-sky average in the detector's reference frame. Through coordinate transformations, we have calculated the noise PSD for any TDI combination, as well as the average response functions for various modes under the metric gravity theory. This allows us to derive the expression for the sensitivity function. Here are the analytical expression for tensor modes:

$$\begin{aligned} R(u) =& \frac{2}{4}C_1\left[\tilde{P}_i(u)\right]f_1(u) + C_2\left[\tilde{P}_i(u)\right]f_2(u) \\ &+ \frac{3}{4}C_3\left[\tilde{P}_i(u)\right]f_3(u) - \frac{3}{4}C_4\left[\tilde{P}_i(u)\right]f_4(u) \\ &+ \frac{1}{4}C_5\left[\tilde{P}_i(u)\right]f_5(u), \end{aligned} \tag{11.21}$$

where

$$C_3\left[\tilde{P}_i(u)\right] = \sum_{i=1}^{3} \mathrm{Re}\left[\tilde{P}_i\tilde{P}^*_{i+1} + \tilde{P}_i\tilde{P}^*_{(i-1)'}e^{iu}\right],$$
$$C_4\left[\tilde{P}_i(u)\right] = \sum_{i=1}^{3} \mathrm{Im}\left[\tilde{P}_i\tilde{P}^*_{i+1} + \tilde{P}_{i'}\tilde{P}^*_{(i-1)'}e^{iu}\right], \tag{11.22}$$
$$C_5\left[\tilde{P}_i(u)\right] = \sum_{i=1}^{3} \mathrm{Re}\left[\tilde{P}_i\tilde{P}^*_{i'} + \tilde{P}_i\tilde{P}^*_{(i-1)'}\right],$$

and

$$\begin{aligned}
f_1(u) &= \frac{4}{3} - \frac{2}{u^2} + \frac{\sin 2u}{u^3},\\
f_2(u) &= \frac{-u\cos u + \sin u}{u^3} - \frac{\cos u}{3},\\
f_3(u) &= \frac{-5\sin u + 8\sin 2u - 3\sin 3u}{8u} + \mathrm{Ci}3u - 2\mathrm{Ci}2u + \mathrm{Ci}u\\
&\quad - \frac{1}{3}\left(\frac{4 + 9\cos u + 12\cos 2u + \cos 3u}{8u^2}\right) + \log\frac{4}{3} - \frac{5}{18}\\
&\quad + \frac{1}{3}\left(\frac{-5\sin u + 8\sin 2u + 5\sin 3u}{8u^3}\right),\\
f_4(u) &= \frac{-5\cos u + 8\cos 2u - 3\cos 3u}{8u} + 2\mathrm{Si}2u - \mathrm{Si}3u - \mathrm{Si}u\\
&\quad + \frac{1}{3}\left(\frac{9\sin u + 12\sin 2u + \sin 3u}{8u^2}\right)\\
&\quad - \frac{1}{3}\left(\frac{8 + 5\cos u - 8\cos 2u - 5\cos 3u}{8u^3}\right),\\
f_5(u) &= -\log 4 + \frac{7}{6} + \frac{11\sin u - 4\sin 2u}{4u} + 2\left(\mathrm{Ci}2u - \mathrm{Ci}u\right)\\
&\quad - \frac{10 + 5\cos u - 2\cos 2u}{4u^2} + \frac{5\sin u + 4\sin 2u}{4u^3}.
\end{aligned} \tag{11.23}$$

Here, $\mathrm{Si}(z) = \int_0^z \sin t/t\,\mathrm{d}t$ and $\mathrm{Ci}(z) = -\int_z^\infty \cos t/t\,\mathrm{d}t$.

We assess the detector's sensitivity functions by examining the generators of the kernel of a homomorphism map as dictated by the underlying TDI equation. This analysis is based on decomposing an arbitrary second-generation TDI combination into a linear combination of independent first and zeroth-generation generators. When the differences in armlengths are disregarded, it emerges that the sensitivity function can be categorized based on a more compact set of independent generators. Specifically, the equal armlength approximation simplifies the underlying TDI equation as well as the resultant analytic expressions of sensitivity functions. By cataloging all fourty-five different TDI combinations up to sixteen links derived through the geometric TDI approach, we derive eleven distinct elementary sensitivity curves in terms of the pertinent generators. The observed degeneracies in sensitivity curves are explained using algebraic arguments. The derived analytic results are anticipated to be instrumental in characterizing the quality and refining the implementation of specific TDI solutions [21].

11.4 Conclusion

In this paper, we present TDI technique for mitigating laser frequency noise and clock noise. We have formulated constraint equations for various TDI combinations, enabling the determination of optimal TDI configurations. Additionally, we develop a ternary symbolic search approach to identify diverse geometric TDI combinations. We analyze the residual characteristics of clock noise under these geometric TDI combinations and derived geometric patterns for clock noise expressions, providing a foundational framework for the development of a general approach to clock noise suppression. Furthermore, we have extended beyond the limitations of specific TDI combinations by decomposing the noise and response functions of multiple data streams, and we have provided analytical expressions for the sensitivity functions of different TDI combinations.

References

1. M. Tinto, S.V. Dhurandhar, Time-delay interferometry. Living Rev. Rel. **24**(1), 1 (2021)
2. M. Tinto, O. Hartwig, Time-delay interferometry and clock-noise calibration. Phys. Rev. D **98**(4), 042003 (2018)
3. O. Hartwig, J.-B. Bayle, Clock-jitter reduction in lisa time-delay interferometry combinations. Phys. Rev. D **103**(12), 123027 (2021)
4. S.V. Dhurandhar, K. Rajesh Nayak, J.Y. Vinet, Algebraic approach to time-delay data analysis for lisa. Phys. Rev. D **65**(10), 102002 (2002)
5. K. Rajesh Nayak, J.Y. Vinet, Algebraic approach to time-delay data analysis for orbiting lisa. Phys. Rev. D **70**, 102003 (2004)
6. e M. Vallisneri, Geometric time delay interferometry. Phys. Rev. D **72**, 042003 (2005). [Erratum: Phys. Rev. D **76**, 109903 (2007)]
7. M. Muratore, D. Vetrugno, S. Vitale, Revisitation of time delay interferometry combinations that suppress laser noise in lisa. Class. Quant. Grav. **37**(18), 185019 (2020)
8. M. Muratore, D. Vetrugno, S. Vitale, O. Hartwig, Time delay interferometry combinations as instrument noise monitors for lisa. Phys. Rev. D **105**(2), 023009 (2022)
9. P.-P. Wang, W.-L. Qian, Y.-J. Tan, W. Han-Zhong, C.-G. Shao, Geometric approach for the modified second generation time delay interferometry. Phys. Rev. D **106**(2), 024003 (2022)
10. M. Tinto, J.W. Armstrong, Cancellation of laser noise in an unequal-arm interferometer detector of gravitational radiation. Phys. Rev. D **59**, 102003 (1999)
11. J.W. Armstrong, F.B. Estabrook, M. Tinto, Time-delay interferometry for space-based gravitational wave searches. Astrophys. J. **527**(2), 814–826 (1999)
12. M. Tinto, J.W. Armstrong, F.B. Estabrook, Discriminating a gravitational wave background from instrumental noise in the lisa detector. Phys. Rev. D **63**, 021101 (2001)

13. C.J. Hogan, P.L. Bender, Estimating stochastic gravitational wave backgrounds with sagnac calibration. Phys. Rev. D **64**, 062002 (2001)
14. J.W. Armstrong, F.B. Estabrook, M. Tinto, Time delay interferometry. Class. Quant. Grav. **20**, S283–S289 (2003)
15. N.J. Cornish, R.W. Hellings, The effects of orbital motion on lisa time delay interferometry. Class. Quant. Grav. **20**, 4851–4860 (2003)
16. M. Vallisneri, Geometric time delay interferometry. Phys. Rev. D **72**, 042003 (2005). [Erratum: Phys. Rev. D **76**, 109903 (2007)]
17. R.W. Hellings, Elimination of clock jitter noise in spaceborne laser interferometers. Phys. Rev. D **64**, 022002 (2001)
18. W. Zhang-Qi, P.-P. Wang, W.-L. Qian, C.-G. Shao, A combinatorial algebraic approach for the modified second-generation time-delay interferometry. Phys. Rev. D **107**, 024042 (2023)
19. W. Zhang-Qi, P.-P. Wang, W.-L. Qian, W.-S. Huang, Y.-J. Tan, C.-G. Shao, Extended combinatorial algebraic approach for the second-generation time-delay interferometry. Phys. Rev. D **108**, 082002 (2023)
20. W.-L. Qian, P.-P. Wang, W. Zhang-Qi, C.-G. Shao, B. Wang, R.-H. Yue, On second-order combinatorial algebraic time-delay interferometry. Phys. Rev. D **108**, 022002 (2023)
21. P.-P. Wang, W.-L. Qian, W. Zhang-Qi, H.-K. Chen, W. Huang, W. Han-Zhong, Y.-J. Tan, C.-G. Shao, Sensitivity functions for geometric time-delay interferometry combinations. Phys. Rev. D **108**(4), 044075 (2023)
22. P.-P. Wang, L. Xiao-Yu, X.-L. Zhao, H.-K. Chen, J. Zhou, W. Huang, Y.-J. Tan, W. Han-Zhong, C.-G. Shao, Time delay interferometry combination with zero-response. Results Phys. **58**, 107481 (2024)
23. M. Tinto, F.B. Estabrook, J.W. Armstrong, Time delay interferometry for lisa. Phys. Rev. D **65**, 082003 (2002)
24. G. Heinzel, J. Jos Esteban, S. Barke, M. Otto, Y. Wang, A.F. Garcia, K. Danzmann, Auxiliary functions of the lisa laser link: Ranging, clock noise transfer and data communication. Class. Quant. Grav. **28**, 094008 (2011)
25. M. Otto, G. Heinzel, K. Danzmann, Tdi and clock noise removal for the split interferometry configuration of lisa. Class. Quant. Grav. **29**, 205003 (2012)
26. P.-P. Wang, Y.-J. Tan, W.-L. Qian, C.-G. Shao, Refined clock-jitter reduction in the sagnac-type time-delay interferometry combinations. Phys. Rev. D **104**(8), 082002 (2021)
27. L. Xiao-Yu, Y.-J. Tan, C.-G. Shao, Sensitivity functions for space-borne gravitational wave detectors. Phys. Rev. D **100**(4), 044042 (2019)
28. C. Zhang, Q. Gao, Y. Gong, B. Wang, A.J. Weinstein, C. Zhang, Full analytical formulas for frequency response of space-based gravitational wave detectors. Phys. Rev. D **101**(12), 124027 (2020)
29. S.L. Larson, W.A. Hiscock, R.W. Hellings, Sensitivity curves for spaceborne gravitational wave interferometers. Phys. Rev. D **62**, 062001 (2000)
30. M. Vallisneri, J. Crowder, M. Tinto, Sensitivity and parameter-estimation precision for alternate lisa configurations. Class. Quant. Grav. **25**, 065005 (2008)
31. N.J. Cornish, L.J. Rubbo, The lisa response function. Phys. Rev. D **67**, 022001 (2003). [Erratum: Phys. Rev. D **67**, 029905 (2003)]
32. M. Vallisneri, C.R. Galley, Non-sky-averaged sensitivity curves for space-based gravitational-wave observatories. Class. Quant. Grav. **29**, 124015 (2012)

Part III
Key Technologies for TianQin

Chapter 12
Research Progress of Charge Management System for TianQin Project

Wei Hong, Bingxue Chen, Liangyu Chu, Honggang Li, Yanzheng Bai, and Ze-Bing Zhou

Abstract The test mass (TM) in space-based gravitational wave detectors acts as an inertial reference but is exposed to high-energy particle radiation, causing charging that interferes with detection. Therefore, charge management for the TM is necessary. This paper presents key advancements in the charge management system for the TianQin project. First, an electrostatic force model is established to determine the TM's charge threshold, which is limited to less than 2×10^{-13} C. And then, a charge management structure is designed using a UV LED light source, and an engineering prototype is developed based on this design. Furthermore, a ground testing system is constructed with a torsion pendulum to evaluate the performance of charge management. The experimental results demonstrate that the resolution of charge measurement is better than 2×10^{-14} C at 1 mHz, and the resolution of charge control is about 6×10^{-14} C, which probide a solid foundation for TianQin project.

12.1 Introduction

In 2015, the Laser Interferometer Gravitational-Wave Observatory (LIGO) first detected gravitational waves, opening a new era for human exploration of the universe [1]. Ground-based gravitational wave detectors face significant challenges, including limitations in arm length limitations, ground vibration noise, and surface gravity gradient variations, which hinder the detection of gravitational waves below 1 Hz. To overcome these limitations, space-based gravitational wave detectors have constituted as optimal program for observing gravitational waves in the millihertz band [2]. These space-based detectors consist of three spacecraft arranged in a triangular constellation, forming a large laser interferometer for gravitational wave detection [3]. The inertial reference of the laser interferometer is the free-floating test mass

W. Hong (✉) · B. Chen · L. Chu · H. Li · Y. Bai · Z.-B. Zhou
National Gravitation Laboratory, MOE Key Laboratory of Fundamental Physical Quantities Measurement and School of Physics, Huazhong University of Science and Technology, Wuhan, China
e-mail: hongwei@hust.edu.cn

J. Luo et al. (eds.), *Proceedings of the 7th International Workshop on the TianQin Science Mission; 25–26 April 2024, Hong Kong SAR, China*, Springer Proceedings in Physics 428, https://doi.org/10.1007/978-981-96-7803-7_12

(TM) inside the spacecraft. Ideally, these TMs perform geodesic motion under the gravitational field alone and therefore provide a true reflection of gravitational wave information. However, high-energetic charged particles in space, such as galactic cosmic rays and solar energetic particles, can cause the charging of the TM. The charged TM interacts with the surrounding electromagnetic field, introducing additional acceleration noise [4], so it is necessary to control the charge on the TM [5].

So far, there are two mature methods for charge control. on method is to use in gravity satellites such as GOCE, CHAMP and GRACE, employs a gold wire to connect the TM and the power ground to neutralize the charge. Due to the thermal noise of the gold wire, the acceleration noise of the TM is limited in the order of 10^{-12} m/s^2/Hz$^{1/2}$, which obviously cannot meet the detection needs of gravitational waves [6]. The other method is a non-contact charge control scheme utilized by GP-B and LISA Pathfinder, using UV light to generate photoelectrons that neutralize the charge. This method theoretically ensures that the acceleration noise of the TM is on the order of 10^{-15} m/s^2/Hz$^{1/2}$, making it a feasible option for gravitational wave detection [7]. Therefore, the second method is used for the charge management of the TianQin project. The detailed process is to first apply modulation voltages on the electrodes that the charged TM will move under the modulation, and the amplitude of the movement is proportional to the amount of the charge, so the charge can be calculated by the data of the movement. And then, by controlling the lighting direction, intensity and duration, the charge on the TM can be controlled to the expected value [8].

TianQin project is a Chinese space-based gravitational wave detection program scheduled for launch around 2035 [9]. Due to shorter Length of interference arm, TianQin requires the charge on the TM more stringent than that of LISA. Therefore, it is necessary to further study the charge management system for TianQin. The subsequent sections are organized as follows: Sect. 12.2 analyses the requirement of charge management system for TianQin, Sect. 12.3 describes the design of charge management system. Section 12.4 tests the performance of the charge management system. Section 12.5 is the conclusion.

12.2 Requirement Analysis

The distribution of the TM and the four electrodes along the X-axis is shown in Fig. 12.1. The distance between the electrodes and the TM is denoted as d_0, and the distance between the centers of electrodes on the same side is b. When the TM is displaced by x_0 along the X-axis (as shown by the dotted line in Fig. 12.1), the electrostatic force on the TM along the X-axis can be obtained using the energy storage equation [10]:

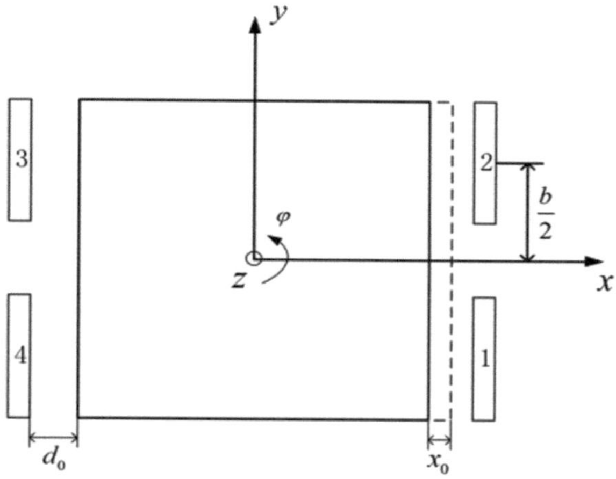

Fig. 12.1 Schematic of the TM with 4 electrodes along the x-axis

$$F_x = \frac{1}{2}\sum_i \frac{\partial C_i}{\partial x}(V_i - V_{\mathrm{TM}})^2 \tag{12.1}$$

where V_i is the potential on the ith electrode, which mainly contains the 100 Hz control voltage $V_{i,AC}$, 100 kHz injection voltage, and the stray potential $V_{i,stray}$. The potential of the TM primarily originates from the TM's charge and the surrounding electric field, denoted as:

$$V_{\mathrm{TM}} = \frac{Q + \sum C_j V_j}{C_{\mathrm{tot}}} \tag{12.2}$$

Therefore, the acceleration of the TM is:

$$a_x = \frac{1}{2m}\sum_i \frac{\partial C_i}{\partial x}\left(V_i - \left(\frac{Q + \sum C_j V_j}{C_{\mathrm{tot}}}\right)\right)^2 \tag{12.3}$$

The sources of acceleration noise are displacement noise $S_x^{1/2}$, charge noise $S_Q^{1/2}$, and voltage noise $S_{V_i}^{1/2}$, so the total acceleration noise is:

$$S_{a_x}^{1/2} = \sqrt{S_x(\frac{\partial a_x}{\partial x})^2 + S_Q(\frac{\partial a_x}{\partial Q})^2 + \sum_{i=1} S_{V_i}(\frac{\partial a_x}{\partial V_i})^2} \tag{12.4}$$

According to the relevant parameters of the TianQin's inertial sensor, the magnitude of electrostatic acceleration noise is shown in Fig. 12.2 [10], where the electrostatic noise curve is plotted at the TM's charge of 3.4×10^{-13} C and the stray potential controlled to 10 mV, the acceleration noise is below 3×10^{-16} m/s^2/Hz$^{1/2}$. The noise terms associated with the TM's charge are mainly the stray potential noise and charge noise, so the acceleration noise introduced by the charge is closely related

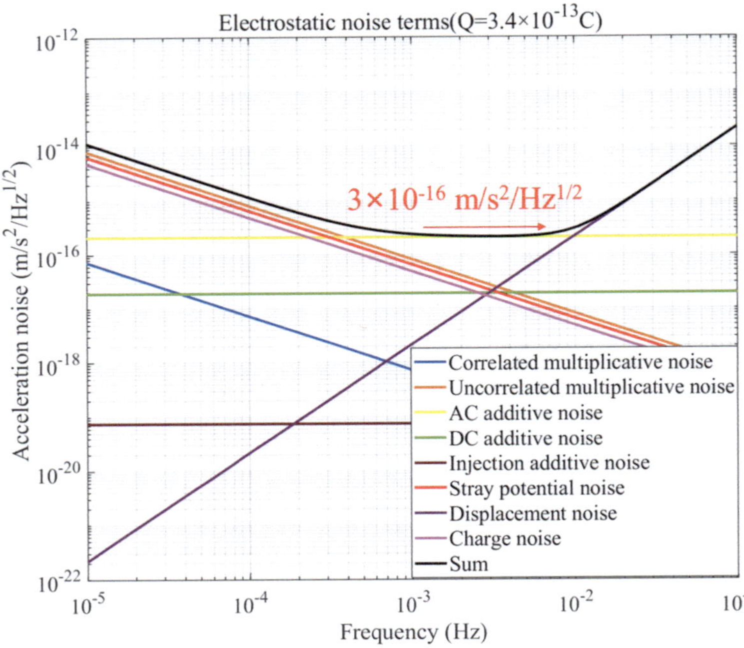

Fig. 12.2 Comparison of electrostatic noise terms

to the stray potential. The electrostatic noise requirement for the TianQin project is 2.4×10^{-16} m/s^2/Hz$^{1/2}$. Considering that the charge value that need to be controlled are sufficiently safe, and allowing some margin for control of the stray potential. Therefore the charge value is further reduced, and it is recommended that the TM's charge be controlled to less than 2.0×10^{-13} C [11].

12.3 System Design

As UV LED has obvious advantages over mercury lamps in terms of response speed, quality, and power consumption, which is suitable as UV light sources for charge management. A structure of charge management system is proposed, as shown in Fig. 12.3. It consists of a measurement and control unit, UV light source, optical fiber, and feedthrough [12]. The system measures charge by applying a modulation voltage on the electrode, which causes a displacement in the TM. This displacement is then recorded and used to calculate the charge. Based on the charge measurement value and a preset charge control strategy, the UV LED is controlled to illuminate the TM or the electrodes and housing through the optical fiber and the feedthrough. The performance of the system must meet the following indicators: the charge needs to be maintained within 2×10^{-13} C in the scientific phase, the maximum discharge

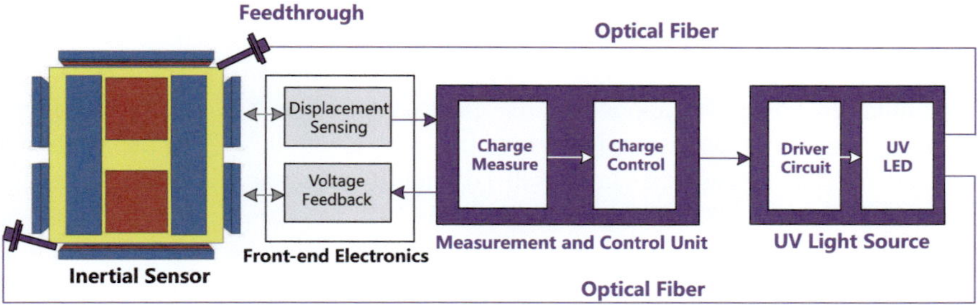

Fig. 12.3 Structural diagram of the charge management system

rate is greater than 10^5 e/s for extreme solar storms, and the discharge resolution superior is to 100 e/s for continuous.

To evaluate the performance of the charge management system, a ground-based test system is built using a vacuum torsion pendulum device and the charge management device, as shown in Fig. 12.4. The model number of the UV light source used in the Charge Management System is Crystal IS OPTAN-255 J-BL, which emits UV light at a wavelength of 255 nm. The entire vacuum cavity is placed on the vibration isolation foundation, and an insulation film is installed outside to mitigate external temperature fluctuation The torsion pendulum system, which includes inertial sensors, is placed inside the vacuum cavity [13].

Fig. 12.4 Charge management test system

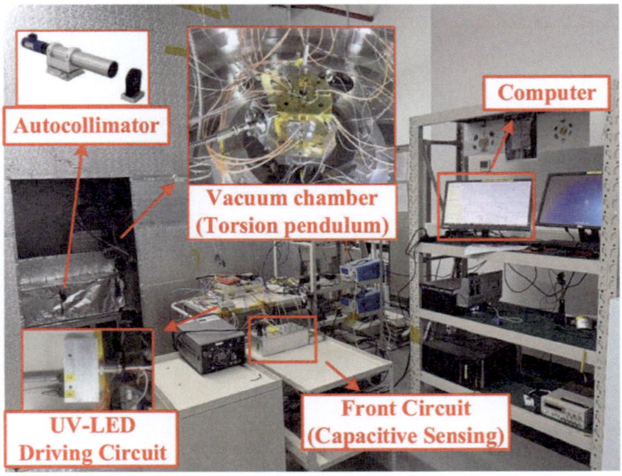

12.4 Performance Testing

A modulation voltage is applied to the electrodes of the X-axis, and the high-frequency carrier voltage is applied to the injection electrodes. Then the rotational angle of the TM is measured using a self-collimator and capacitive sensing, respectively, in which the TM's charge is estimated [14]. To evaluate the resolution of the two measurement methods, firstly six sets of experiments with a modulation voltage amplitude of 2 V and modulation voltage frequency of 0.3, 1, 3, 10, 30 and 100 mHz are carried out, and compared with the theoretical precision results, as shown in Fig. 5a. And then five sets of experiments with modulation amplitudes of 0.2, 0.5, 1, 1.5 and 2 V at a modulation frequency of 10 mHz are carried out and the results are shown in Fig. 5b. The measurement resolution of two measurement methods are both achieved 2×10^{-14} C at 1 mHz, which provides a foundation for the charge measurement on the non-sensitive axis to avoid interfering with scientific measurement [13].

The change curve of TM's potential with discharging/charging time can be obtained by charge control experiments. Also define the increase of the TM's charge as charging and vice versa as discharging. The charge control experiments, a modulating voltage with a frequency of 10 mHz and an amplitude of 2 V is first applied to the four control electrodes in the X-axis direction for the measurement of the TM's potential. After about 30 min of measurement, the modulating voltage is disconnected, and a UV light with an optical power of about 50 nW to illuminate the TM or electrodes and housing for a duration of 300 s. This is repeated until the TM's potential is close to saturation. The results of the charging and discharging process for TM as shown in Fig. 12.6. The blue points are the measured values of the TM's potential, and the red curve is the fitting result of the measured data. The TM's potential changes exponential over time, and the related study provided the following expression [15]:

$$V_{tm}(t) = \left(V_0 - V_{TMeq}\right)e^{-\frac{t}{\tau}} + V_{TMeq} \quad (12.5)$$

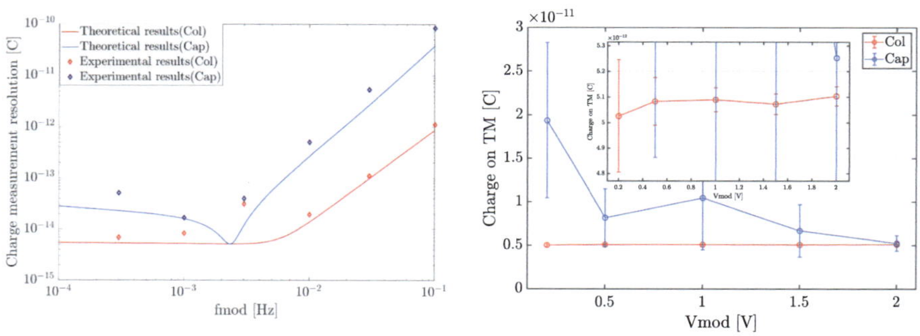

(a) (b)

Fig. 12.5 Experimental results of charge measurement. **a** Different modulation frequencies; **b** Different modulation amplitudes

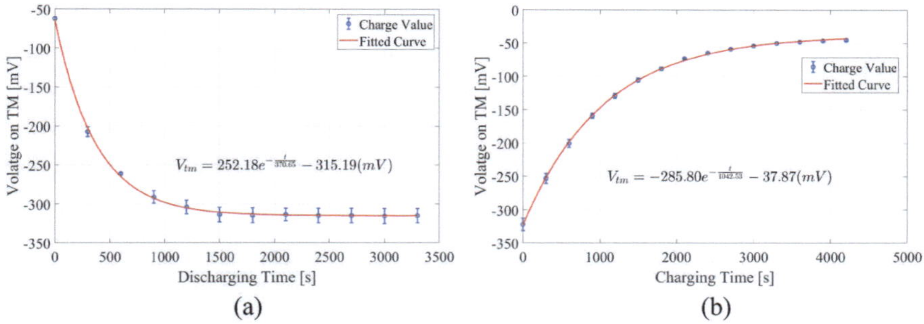

Fig. 12.6 Charge/discharge modelling for test mass. **a** Discharging; **b** Charging

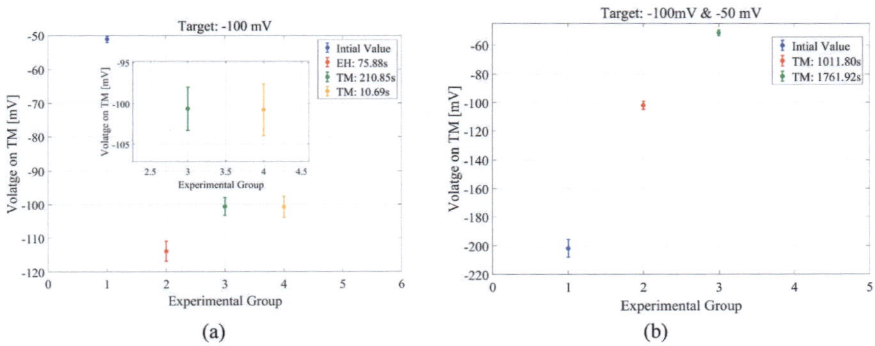

Fig. 12.7 Effect of charge control. **a** Target: −100 mV; **b** Target: −100 mV and −50 mV

where V_0 is the initial potential of the TM, V_{TMeq} is the equilibrium potential of the TM, and τ is the time constant related to the optical power.

Charge control performance is evaluated based on the fitted model. Specifically, after measuring the current potential value of the TM, the charging and discharging time is determined based on the target potential value and the fitted model at an optical power of 50 nW. Figure 7a shows the control effect with the initial potential of about −50 mV and the target potential of −100 mV, and the labels are the required charging/discharging times. Figure 7b shows the control effect with the initial potentials of about −200 mV and target potentials of −100 and −50 mV. The experimental results show that the resolution of charge control is better than 2 mV, i.e., 6×10^{-14} C.

12.5 Conclusion

This paper introduces the main progress of the charge management system of TianQin project. According to the electrostatic noise requirement, the TM's charge is required less than 2×10^{-13} C. And then, a charge management system structure based on a UV LED light source is designed, and an engineering prototype based on the structure is developed. Finally, the performance of the charge management system is experimentally verified by using the ground testing system. The experimental results show that the measurement resolution is better than 2×10^{-14} C at 1 mHz, and the charge control resolution is approximately 6×10^{-14} C. In the future, we will focus on improving the prototype's reliability.

References

1. C. Cahillane, G. Mansell, Review of the advanced LIGO gravitational wave observatories leading to observing run four. Galaxies 10 (2022). https://doi.org/10.3390/galaxies10010036
2. E. Barausse, E. Berti, T. Hertog et al., Prospects for fundamental physics with LISA. Gen. Relativ. Gravit. **52**, 1–33 (2020)
3. P. Amaro-Seoane, H. Audley, S. Babak et al., Laser interferometer space antenna. arXiv preprint arXiv:170200786 (2017)
4. C. Lei, W. Su, W. Hong et al., Simulation for the test mass charging rate in the Tianqin orbit. Cl. Quantum Grav. **41**, 025001 (2024). https://doi.org/10.1088/1361-6382/ad105a
5. M. Armano, H. Audley, J. Baird et al., Charging of free-falling test masses in orbit due to cosmic rays: results from LISA Pathfinder. Phys. Rev. D **107**, 062007 (2023). https://doi.org/10.1103/PhysRevD.107.062007
6. L. Liu, Y. Bai, Z. Zhou et al., Measurement of the effect of a thin discharging wire for an electrostatic inertial sensor with a high-quality-factor pendulum. Cl. Quantum Grav. **29**, 055010 (2012)
7. M. Armano, H. Audley, G. Auger et al., Sub-femto- g free fall for space-based gravitational wave observatories: LISA pathfinder results. Phys. Rev. Lett. **116**, 231101 (2016). https://doi.org/10.1103/PhysRevLett.116.231101
8. S. Buchman, T. Quinn, G. Keiser et al., Charge measurement and control for the gravity probe B gyroscopes. Rev. Sci. Instrum. **66**, 120–129 (1995)
9. Y. Gong, J. Luo, B. Wang, Concepts and status of Chinese space gravitational wave detection projects. Nat. Astron. **5**, 881–889 (2021). https://doi.org/10.1038/s41550-021-01480-3
10. T.J. Sumner, G. Mueller, J.W. Conklin et al., Charge induced acceleration noise in the LISA gravitational reference sensor. Cl. Quantum Grav **37**, 045010 (2020). https://doi.org/10.1088/1361-6382/ab5f6e
11. J. Luo, L.-S. Chen, H.-Z. Duan et al., TianQin: a space-borne gravitational wave detector. Cl. Quantum Grav. **33**, 035010 (2016). https://doi.org/10.1088/0264-9381/33/3/035010
12. F. Yang, Y. Bai, W. Hong et al., Investigation of charge management using UV LED device with a torsion pendulum for TianQin. Cl. Quantum Grav. **37**, 115005 (2020). https://doi.org/10.1088/1361-6382/ab8489
13. H. Li, G. Li, W. Hong et al., Capacitive sensing based charge measurement for space inertial sensors. IEEE Trans. Instrum. Meas. (2024)
14. B. Chen, W. Hong, H. Li et al., Using finite element simulation to evaluate charge measurement precision for space inertial sensors. Meas. Sci. Technol. **35**, 045026 (2024)
15. S. Apple, A.D. Álvarez, S.P. Kenyon et al., Design and performance characterization of a new LISA-like (laser interferometer space antenna-like) gravitational reference sensor and torsion pendulum testbed. Rev. Sci. Instrum. **94**, 054502 (2023). https://doi.org/10.1063/5.0135125

Chapter 13
The Simulation for the AC Magnetic Susceptibility of Test Mass in TianQin

Jia-Qi Wu, Xue-Yuan Li, Dan Peng, and Qi Liu

Abstract Aiming to detect gravitational wave signals at the frequency band of 0.1mHz−1Hz in space, TianQin has the requirement of the residual acceleration noise of the test mass which should be lower than $10^{-15}\text{ms}^{-2}\text{Hz}^{-1/2}$, thus all influence factors acting upon the test mass should be restricted within this requirement. In this paper, We used simulation methods to explore the eddy current effects of the test mass in TianQin, and the processes were presented in detail. The simulation results were compared with both analytical solution and measurement result and they show an excellent agreement.

13.1 Introduction

As a significant Chinese space-based gravitational wave detector, TianQin aims to detect the gravitational wave signals in the frequency range of 0.1mHz−1Hz [1] by the tiny displacement change between the test masses loaded on TianQin satellites.TianQin has a strict requirement on the residual acceleration noise of test masses at $1 \times 10^{-15}\text{ms}^{-2}\text{Hz}^{-1/2}$ [2], thus all tiny noises should be taken into account.

The magnetic disturbance is expected to contribute 24% of TianQin's error budget of residual acceleration noise, therefore, it's significant for us to explore the influence of magnetic disturbance [1, 3, 4]. Usually, the magnetic force is estimated by the coupling of remnant magnetic dipole m_r and susceptibility χ from the material of the test mass to the ambient magnetic field. This magnetic effect has been investigated by many researchers [4–6]. However, the eddy current effects in the test mass have seldom been discussed in previous work. The eddy current is caused by time-varying magnetic fields, while the magnetic field changes, which will induce the eddy current in the test mass. It's hard for us to calculate the magnetic force caused by the eddy current because of the unknown total magnetic field's distribution and the weak

J.-Q. Wu · X.-Y. Li · D. Peng · Q. Liu (✉)
TianQin Research Center for Gravitational Physics and School of Physics and Astronomy, Sun Yat-sen University (Zhuhai Campus), Zhuhai, People's Republic of China
e-mail: liuq239@sysu.edu.cn

J. Luo et al. (eds.), *Proceedings of the 7th International Workshop on the TianQin Science Mission; 25–26 April 2024, Hong Kong SAR, China*, Springer Proceedings in Physics 428, https://doi.org/10.1007/978-981-96-7803-7_13

induced eddy current. Thus, we will discuss the eddy current effects of the test mass in TianQin.

13.2 Eddy Current Calculation

Treat the test mass as a magnetic dipole $\boldsymbol{m}$ induced by an external magnetic field, the magnetic force applied on the test mass can be written as [7]

$$\boldsymbol{F} = \int_v (\boldsymbol{m} \cdot \nabla \boldsymbol{B}) dV = \frac{1}{\mu} \int_v (\chi_e \boldsymbol{B} \cdot \nabla \boldsymbol{B}) dV, \tag{13.1}$$

here χ_e is the magnetic susceptibility of the test mass caused by eddy current caused by the ac magnetic field, which can be calculated by the induce eddy current and the total magnetic field (see Eq.13.4). Because $\boldsymbol{m}$ is related to the current inside the test mass, we can calculate the force induced by eddy current $\boldsymbol{J}$ using the concept of magnetic susceptibility χ_e caused by eddy current $\boldsymbol{J}$. However, it is difficult for us to get the numerical solution of $\boldsymbol{J}$ when the test mass is a cube. Thus, we decide to use simulation methods to calculate magnetic susceptibility χ_e of TianQin's test mass to explore the eddy current effects.

13.3 Simulation Processes

13.3.1 Simulation Principle

The simulation software we use here is COMSOL. We build the model of the test mass and locate it in cartesian coordinates. A uniform magnetic field $\boldsymbol{B} = B_0 e^{iwt}$ is applied in the z direction (Fig. 13.1b, c). The simulation principle is as Fig. 13.1a.

After the simulation, we can get the eddy current of the test mass. According to equation

$$\boldsymbol{m} = \frac{1}{2} \int_v \boldsymbol{r} \times \boldsymbol{J} dV, \tag{13.2}$$

we have the volume-averaged magnetization

$$\boldsymbol{m}_v = \frac{1}{V} \int_V \boldsymbol{r} \times \frac{\boldsymbol{J}}{2} dV, \tag{13.3}$$

and get the magnetic susceptibility χ_e contributed by eddy current:

$$\chi_e = \frac{\boldsymbol{m}_v}{\boldsymbol{H}}. \tag{13.4}$$

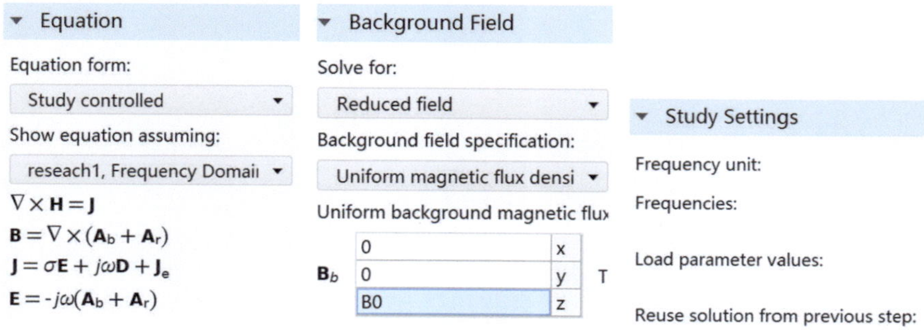

(a) (b) (c)

Fig. 13.1 **a** Simulation Principle. **b** 2-cm-diameter copper sphere in a uniform magnetic field. **c** The simulation result of copper sphere's χ_e and the calculation result of L. D. Landau

13.3.2 Verification of Simulation

Before we simulate the Au-Pt test mass of TianQin, the verification of the simulation method is necessary.

Firstly, we build a 2-cm-diameter copper sphere and mesh it (Fig. 13.2a). Then, following the steps mentioned in Sect. 13.3.1, we can get the magnetic susceptibility χ_e of the copper sphere, and compare it with the calculation result of L. D. Landau [8] (Fig. 13.2c).

We also simulate the situation when the model is a 2-cm-side copper cube (Fig. 13.3a). The simulation result is compared with measurement result [9] (Fig. 13.3c)

These two comparisons show that the simulation method is accurate in estimating the eddy current at the uniform external magnetic field. Therefore, exploring the eddy current effect by the simulation method is available.

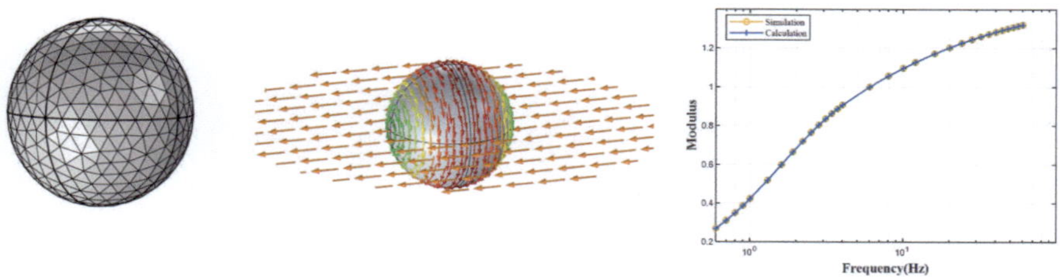

(a) (b) (c)

Fig. 13.2 **a** 2-cm-diameter copper sphere after Finite Element Mesh Refinement. **b** 2-cm-diameter copper sphere in a uniform magnetic field. **c** The simulation result of copper sphere's χ_e and the calculation result of L. D. Landau

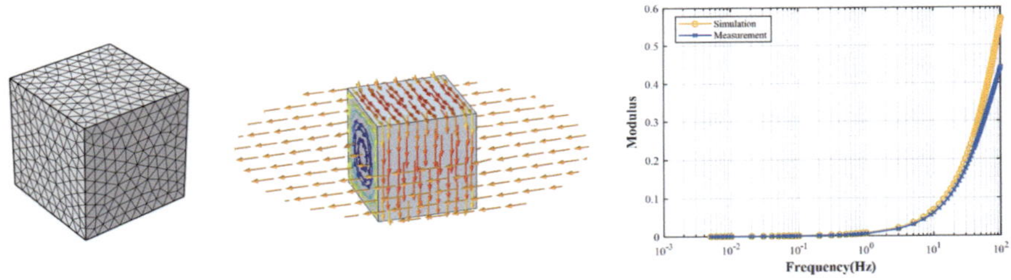

(a) (b) (c)

Fig. 13.3 **a** 2-cm-side copper cube after Finite Element Mesh Refinement. **b** 2-cm-side copper cube in a uniform magnetic field. **c** The simulation result of copper cube's χ_e and the measurement result

13.3.3 Simulation of the Au-Pt Test Mass

The test mass carried on the TianQin satellite is like a 5-cm-side Au-Pt cube (Fig. 13.4a) whose weight is about 2.5 kg. We use the simulation method to estimate test mass's χ_e and discuss the influence caused by different directions of the external magnetic field, the results are shown in Fig. 13.4c. Through the simulation result, we can know that the direction of the magnetic field has little impact on χ_e, the relative errors caused by it are all in –1%–2.5%.

13.4 Summary

In this paper, we use the simulation method to explore the eddy current effect of the test mass in TianQin. The accuracy of the simulation method has been verified through the comparisons with measurement data and calculation results. By this method, we successfully obtain the test mass's χ_e.

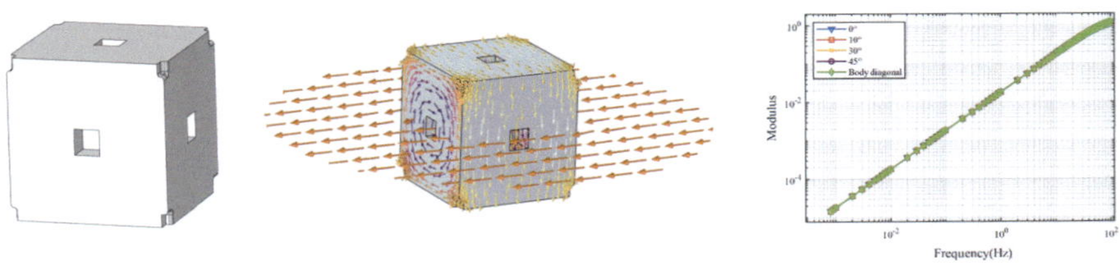

(a) (b) (c)

Fig. 13.4 **a** Test mass model for TianQin. **b** Test mass model for TianQin in a uniform magnetic field. **c** Modulus of TMs' χ_e estimated by simulation

References

1. J. Luo, L.S. Chen, H.Z. Duan, Y.G. Gong, S. Hu, J. Ji, Q. Liu, J. Mei, V. Milyukov, M. Sazhin, C.G. Shao, V.T. Toth, H.B. Tu, Y. Wang, Y. Wang, H.C. Yeh, M.S. Zhan, Y. Zhang, V. Zharov, Z.B. Zhou, Classical Quant. Gravity **33** 035010 (2016). https://dx.doi.org/10.1088/0264-9381/33/3/035010
2. J. Mei, Y. Z. Bai, J. Bao, et al., *Progress of Theoretical and Experimental Physics* (2021) ISSN 2050-3911 https://doi.org/10.1093/ptep/ptaa114
3. W. Su, Y. Wang, Z. B. Zhou, et al., Classical Quant. Gravity **37**, 185017 (2020). https://dx.doi.org/10.1088/1361-6382/aba181
4. H. Yin, D. Y. Tan, M. Hu, et al., Phys. Rev. Appl. **15**(1), 014008 (2021). https://link.aps.org/doi/10.1103/PhysRevApplied.15.014008
5. M. Diaz-Aguiló, E. García-Berro, A. Lobo, Phys. Rev. D **85**(4), 042004 (2012). https://link.aps.org/doi/10.1103/PhysRevD.85.042004
6. W. Su, Z. B. Zhou, Y. Wang, et al., Phys. Rev. D **108**(10), 103030 (2023). https://link.aps.org/doi/10.1103/PhysRevD.108.103030
7. M. Armano, H. Audley, J. Baird, et al., Mon. Notices Royal Astron. Soc. **494**, 3014–3027 (2020). ISSN 0035-8711 (*Preprint* https://academic.oup.com/mnras/article-pdf/494/2/3014/33129159/staa830.pdf) https://doi.org/10.1093/mnras/staa830
8. L. Landau, E. Lifshitz, *Chapter VII–quasi-static Electromagnetic Field Electrodynamics of Continuous Media (Second Edition)* (*Course of Theoretical Physics* vol 8) ed LANDAU L and LIFSHITZ E (Pergamon, Amsterdam, 1984), pp. 199–224 second edition ed ISBN 978-0-08-030275-1. https://www.sciencedirect.com/science/article/pii/B9780080302751500138
9. J.H. Xu, Q. Liu, X. Luo, et al., Phys. Rev. Appl. **18**(4), 044010 (2022). https://link.aps.org/doi/10.1103/PhysRevApplied.18.044010

Chapter 14
Preliminary Research on TianQin-1 In-Orbit Residual Acceleration Caused by the Cold Gas Attitude Control Thrusters

Zhengyuan Shen, Lin Cai, and Ze-Bing Zhou

Abstract TQ-1 satellite is the first technology demonstration satellite of the TianQin project, which is equipped with a high-precision space inertial sensor. In the accelerometer mode, the inertial sensor can measure the non-gravitational forces on the TQ-1 satellite. However, there are many spikes in the inertial sensor measurement. We found that the spikes are mainly caused by the cold gas attitude control thruster, so we try to build physical models to explain these spikes. To begin with, a statistical analysis is used to recover the shape characteristics of the thruster spikes in 3 axes. and then, we introduced an accelerometer response model and an attitude dynamics model that perfectly fits the shape of the thruster spikes. What's more, by using this model, we can accurately estimate the installation deviation of the thrusters and the center of mass deviation between the inertial sensor and the satellite.

14.1 Introduction

Facing the demand for space gravitational-wave (GW) detection, Luo Jun's team proposed the first domestic space-based GW detection mission "TianQin Progect" in 2014 [1]. An equilateral triangle formation is constructed in the geocentric orbit at an altitude of 10^5 m for detecting GWs signals in the frequency band of 10^{-4} $\sim$ 0.1Hz. "TianQin-1 (TQ-1) is the first technology demonstration satellite of the TianQin Project [2], which was successfully launched on December 20, 2019, from Taiyuan Satellite Launch Center.

Z. Shen · L. Cai (✉) · Z.-B. Zhou
National Gravitation Laboratory, MOE Key Laboratory of Fundamental Physical Quantities Measurement, and School of Physics, Huazhong University of Science and Technology, Wuhan, China
e-mail: cailin@hust.edu.cn

J. Luo et al. (eds.), *Proceedings of the 7th International Workshop on the TianQin Science Mission; 25–26 April 2024, Hong Kong SAR, China*, Springer Proceedings in Physics 428, https://doi.org/10.1007/978-981-96-7803-7_14

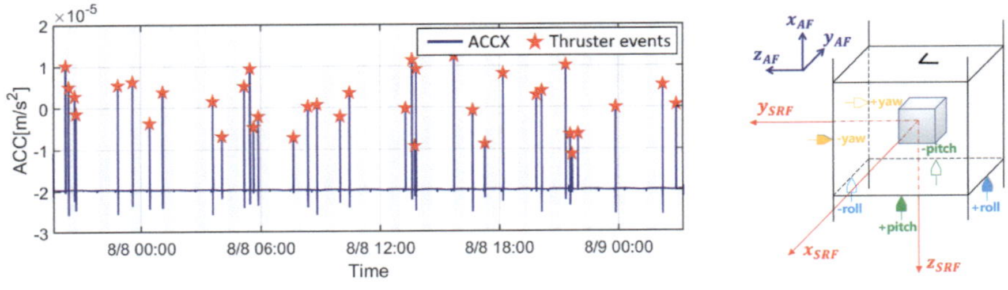

Fig. 14.1 Thruster spikes in linear acceleration (along-track component) (Left) & Attitude control thrusters onboard TQ-1 satellite (Right)

The inertial sensor at the center of the TQ-1 is designed to measure the displacements of the test mass (TM) or the non-gravitational accelerations acting on the satellite surface. TQ-1 is in a sun-synchronous orbit with an orbital altitude of about 628 km. At this orbital altitude, the non-gravitational forces on the satellite are mainly atmospheric drag, solar radiation and albedo. Unfortunately, while measuring the non-gravitational forces on the satellite, the inertial sensor also measured various disturbances generated by the satellite platform, among which the firing of the attitude-controlled cold gas thrusters was the most prominent. Figure 14.1 illustrates the existence of thrust spikes in accelerometer measurements.

The TQ-1 attitude control system actuator consists of a magnetic torquer and a cold-gas propulsion system. The magnetic torquer is the main attitude control actuator of the satellite, which generates the magnetic torque required for satellite attitude control by generating magnetic moment and interacting with the geomagnetic field. When the magnetic torquer is unable to provide the magnetic torque required for satellite attitude control, the cold gas thrusters are activated to compensate for the attitude error.

TQ-1's attitude-controlled cold gas propulsion system consists of six 10mN cold gas thrusters. Figure 14.2 gives the layout of the installation of the cold gas thrusters on TQ-1. A cold gas thruster controls a single attitude direction, which means that each attitude control must be accompanied by a linear acceleration rather than a force couple; at the same time, The presence of thruster installation bias and center-of-mass bias contributes residual acceleration to all linear acceleration directions of the inertial sensors, which are either real non-conservative forces or purely interfering signals that need to be differentiated or orbit determination will be affected. Therefore, in this research we try to:

- Build a dynamic model to explain the physical mechanism of thrust spikes.

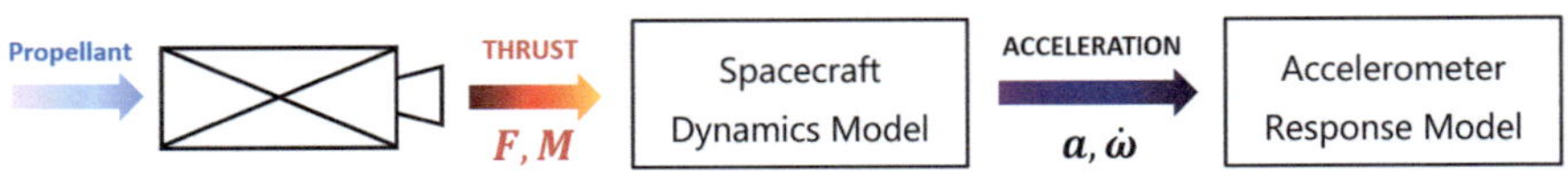

Fig. 14.2 General dynamic model of thrust spikes formation

- Perform statistics on onboard data and fit them by using our dynamic model.
- Accurately estimate the dynamic model parameters.

14.2 Physical Model

Firstly, we will introduce the general dynamic model of thrust spike generation in accelerometer measurements in the Fig. 14.2.

Our model fully accounts for the dynamics from the cold gas propulsion system to spacecraft dynamics modeling to accelerometer response modeling [3].

$$\vec{F} = \dot{m}\overrightarrow{V_e} + A_e(P_e - P_0) \tag{14.1}$$

with

$\dot{m}$ … propellant mass flow rate V_e … propellant exhaust velocity A_e … nozzle exit area P_e … gas pressure at the nozzle exit P_0 … ambient pressure.

It is worth noting that although the pressure difference and the nozzle area are scalar quantities, the resulting pressure thrust is still a vector, and it has a specific direction along the normal to the nozzle's exit plane (perpendicular to the nozzle surface). Since there are installation deviations in the thrusters, once we have the thrust equation, the next step is to analyze the additional residual acceleration in the non-thrust direction caused by the installation deviations.

As is shown in the Fig. 14.3. We consider the contribution of both thruster installation bias and center-of-mass deviation. When there is an angular deviation of the thruster mounting, it will contribute disturbance linear acceleration to the non-thrust direction; when there is a positional deviation of the thruster mounting, it will contribute disturbance moment to the non-posture control direction; and when the center of mass of the satellite deviates from the center of mass of the test mass, the thruster operation at this time produces disturbances in the direction of linear acceleration corresponding to the non-posture control axis.

Unlike the former, the center-of-mass deviation is apparently not really a non-gravitational signal acting on the satellite, but a pure disturbance. By the results of the in-orbit center-of-mass calibration, the center-of-mass bias disturbance present in the thrust spike should be subtracted out, and only the real non-gravitational signal should be retained.

We first give the spacecraft dynamics model below:

$$\overrightarrow{H} + \vec{\omega} \times \vec{H} = \vec{M} = L \times \vec{F} \tag{14.2}$$

where L denotes the length of the force arm, in addition to the length of the force arm designed for ideal conditions, the contribution of the two deviations is considered and is given by the following equation.

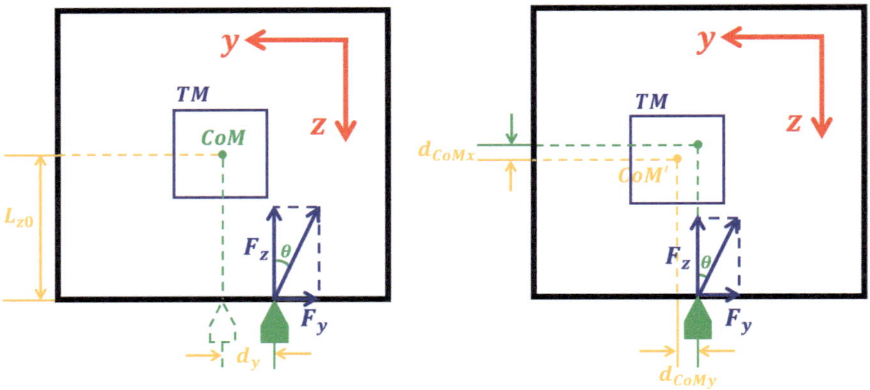

Fig. 14.3 Installation deviation (left) & Center-of-mass deviation (right)

$$L_y = L_{y0} + d_y + d_{CoMy} \tag{14.3}$$

Figure 14.4 gives the accelerometer response model. We use a transfer function model, which is divided into four systems, including inertial sensor heads, capacitive displacement sensors, PID controllers and electrostatic actuators [5].

Based on the above model, we can give an expression for the total transfer function of the inertial sensor in accelerometer mode:

$$H_{OL}(s) = H_m H_S H_C H_a \tag{14.4}$$

$$H_{CL}(S) = \frac{H_{OL}(s)}{1 + H_{OL}(s)} \tag{14.5}$$

With this transfer function expression, we can convert any signal or disturbance input to the inertial sensor into the measured output of the accelerometer.

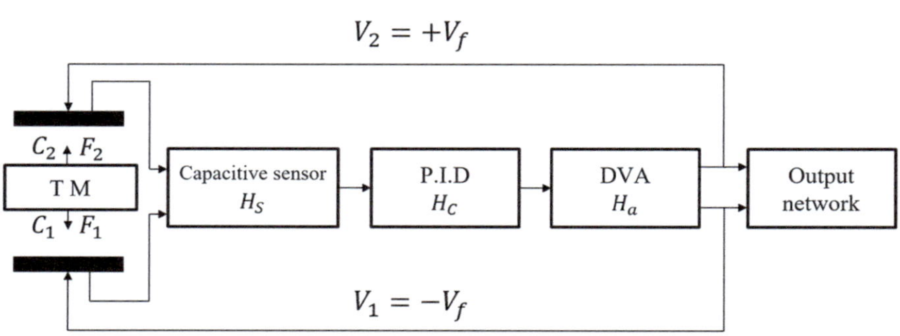

Fig. 14.4 Inertial sensor's response model in Accelerometer Mode

14.3 Result

We begin by counting the responses of the inertial sensors to the thrusters. As shown below, we counted the response spikes of the inertial sensors for the same thrusters working the same pulse widths. From Fig. 14.5, we can find that all six axes of the inertial sensor are present in the single thruster operation with good repeatability, which reflects the stability of the thruster operation as well as the accuracy of the inertial sensor measurements.

A comparison of the statistics for the 3 linear acceleration directions with the response model we have given is given in the Fig. 14.6. The thrust input we use in the system is an approximate square wave signal [4], which is introduced into the processing flow as shown in Fig. 14.3 to obtain our desired response model.

With the inclusion of the accelerometer response model, the thruster events can be perfectly fitted, and the perturbations due to the center of mass shift need to be corrected before the specific parameters of the accelerometers can be obtained, according to Eq. 14.3. The center-of-mass deviation of the satellite, derived from the results of the last calibration prior to statistics, is given in Table 14.1.

Finally, using least squares estimation, we can obtain on-orbit estimates of installation bias as well as total thrust. As shown in Fig. 14.7, the results exhibit good reproducibility, which suggests that these physical quantities are relatively stable over short periods of time.

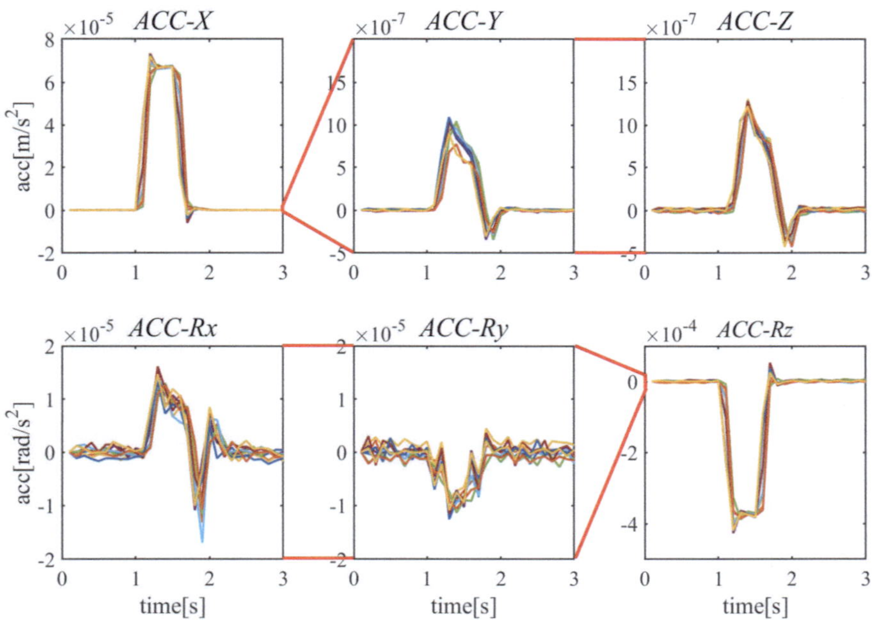

Fig. 14.5 Spikes caused by the—Pitch Thruster

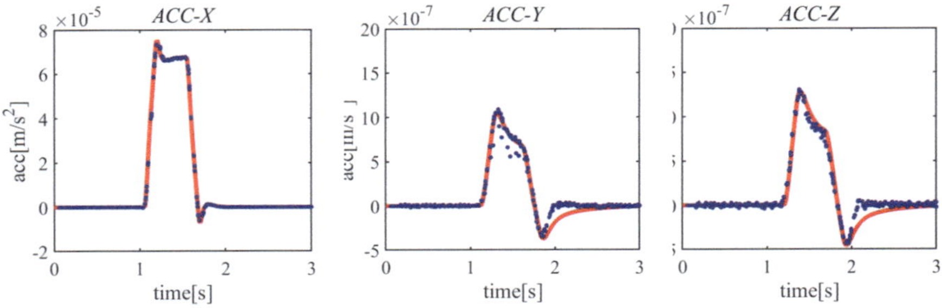

Fig. 14.6 Fitting in-orbit statistics using response models

Table 14.1 Center of mass deviation measurement

Orientation	CoM of gravity deviation (mm)
X	−0.30
Y	0.19
Z	−284.53

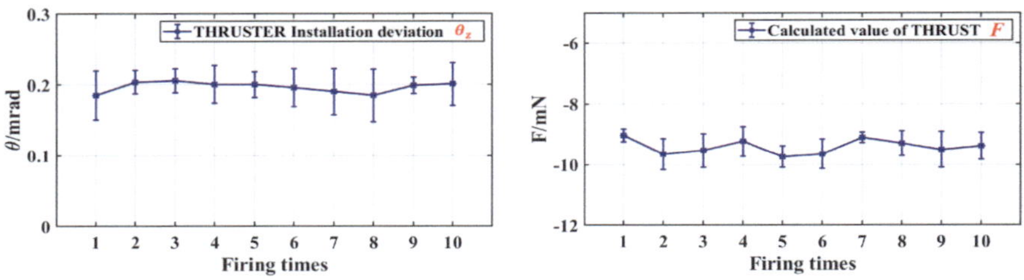

Fig. 14.7 Least square estimation result of angle deviation and total thrust

14.4 Summary and Outlook

This work focuses on the mechanistic modeling of thruster disturbance signals present in inertial sensors. A distinction is made between the pure disturbance of the inertial sensors during the operation of the cold gas thrusters and the real acceleration signal acting on the satellite. In addition to this, this work provides a comprehensive statistic of thrust spikes in orbit and fits it to our given response model, which shows that the established response model is consistent with the measurements of the inertial sensors. After obtaining the fitting results, we used the algorithm of least squares estimation to obtain the results of the angular deviation of the thruster mounting and the actual thrust.

In future work, we will apply this physical model to study the effects of cold gas thruster-generated disturbances on the assessment of inertial sensor on-orbit performance, on satellite orbits, and on gravity field recovery.

References

1. J. Luo, L.S. Chen, H.Z. Duan, Y.G. Gong, S. Hu, J. Ji et al., TianQin: a spaceborne gravitational wave detector. Cl. Quantum Grav. **33**(3), 035010 (2016)
2. J. Luo, Y.Z. Bai, L. Cai, B. Cao, W.M. Chen, Y. Chen et al., The first round result from the TianQin-1 satellite. Cl. Quantum Grav. **37**(18), 185013 (2020)
3. T. Bandikova, The role of attitude determination for inter-satellite ranging. Fachrichtung Geodäsie und Geoinformatik der Leibniz-Universität (2015)
4. T. Bandikova, C. McCullough, G.L. Kruizinga et al., GRACE accelerometer data transplant. Adv. Space Res. **64**(3), 623–644 (2019)
5. A.A. Krasnov, V.G. Rozentsvein, Electrostatic accelerometers for space applications: modern state and prospects of development. Gyroscopy Navig. **13**(2), 59–81 (2022)

Chapter 15
Measurement Scheme of Release Impulse Based on a Compound Pendulum for the TianQin Project

Lingling Yang, Xin Li, Yongqi Liu, Wencan Li, and Qing Li

Abstract In the TianQin project, the caging and release technology of the test mass is one of the key technologies. The test mass will be released to free fall for the science phase in space. The residual release impulse is required to be on the order of magnitude of 1×10^{-5} kg m/s. In this work, the release impulse of the test mass is proposed to be measured with a compound pendulum for the TianQin project, and the expected uncertainty is on the order of magnitude of 1.1×10^{-7} kg m/s. The electrostatic feedback control of the pendulum is used to reduce the nonlinear effect of the pendulum.

15.1 Introduction

The TianQin project is a space-borne gravitational wave detection program independently proposed by China [1], and the inertial sensor is one of the core payloads in this project [2, 3]. During the inertial sensor, the caging and release technology is used to lock the test mass on the ground and release it in space. So far, the caging and release technology was only verified by LISA Pathfinder in space [4, 5]. To test the release impulse on the ground, the University of Trento in Italy has designed the TMMF (Transferred Momentum Measurement Facility) and performed a series of tests [6, 7].

In the TianQin project, the test mass will be an Au-Pt alloy cubic block [8]. A frame of electrodes surrounds the test mass with a capacitive gap of about 4 mm along the sensitive axis. According to the electrostatic control force authority, the maximum release impulse of the test mass is required to be on the order of magnitude of 10^{-5} kg·m/s. To measure the release impulse, our team has carried out a preliminary measurement with the compound pendulum on the ground [9]. This work proposes

L. Yang · X. Li · Y. Liu · W. Li · Q. Li (✉)
MOE Key Laboratory of Fundamental Physical Quantities Measurement,
Hubei Key Laboratory of Gravitation and Quantum Physics, PGMF and School of Physics,
Huazhong University of Science and Technology, Wuhan, People's Republic of China
e-mail: qingli@hust.edu.cn

J. Luo et al. (eds.), *Proceedings of the 7th International Workshop on the TianQin Science Mission; 25–26 April 2024, Hong Kong SAR, China*, Springer Proceedings in Physics 428, https://doi.org/10.1007/978-981-96-7803-7_15

to use the electrostatic feedback control for the compound pendulum to measure the release impulse.

15.2 Design of Measurement System

The measurement system for testing the release impulse with the electrostatic feedback control based on the compound pendulum is shown in Fig. 15.1. The test mass is an aluminum cubic block with dimensions of $5 \times 5 \times 5$ cm^3, which is placed in the middle of the pendulum body. The pendulum is symmetrically suspended by two thin beryllium copper strips. The stiffness of the strip is evaluated to be 9×10^{-4} Nm/rad according to the Young's modulus of the beryllium copper and the dimensions of the strip. The stiffness contributed by the gravity of the pendulum is the main part, which is on the order of magnitude of 0.3 Nm/rad.

The release tips are mounted on opposite sides of the test mass and they are controlled by the six-freedom motion platforms. A reflector is installed on one side of the pendulum body. The ELCOMAT 3000 electronic autocollimator with an accuracy of 0.25" is used to monitor the twist of the pendulum. The capacitive plates are mounted on the top of the pendulum to produce the electrostatic force for the feedback control.

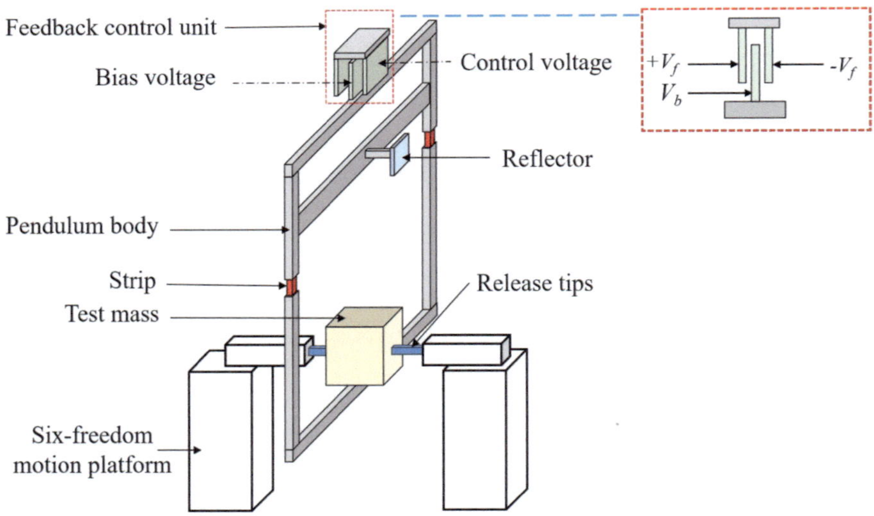

Fig. 15.1 Schematic diagram of the measurement system. The test mass is an aluminum cubic block with dimensions of $5 \times 5 \times 5$ cm^3 and installed on the pendulum. The size of the pendulum frame is $165 \times 100 \times 15$ mm^3. The pendulum is symmetrically suspended by two $1 \times 8 \times 0.02$ mm^3 strips. The test mass is locked and released by the aluminum tips which are installed on the opposite sides. The capacitive plate with dimensions of $40 \times 40 \times 5$ mm^3 are installed on the top of the pendulum

When the release tips are driven by the six degrees of freedom motion platforms, the test mass is locked by them. Then during the release process, if the adhesion force on both sides is not equal, it results in a release impulse I for the pendulum. By now the pendulum acquires an initial angular velocity $\dot{\theta}_0$ and an initial angle θ_0. At this moment, we consider t=0. The solution for the release impulse I can be expressed as [9, 10]:

$$I = \frac{K\theta_r}{b(\omega_n{}^2 + \beta^2)\sqrt{(\frac{1}{\omega_n \tan\varphi + \beta})^2 + (\frac{1}{\omega_n} - \frac{\beta}{\omega_n^2 \tan\varphi + \beta\omega_n})^2}}, \tag{15.1}$$

where K is the stiffness of the pendulum, θ_r and φ are the initial amplitude and initial phase of the pendulum, b is the force arm of the pendulum, ω_n is the oscillation frequency and β is the damping coefficient, respectively.

When the amplitude of the pendulum becomes large, the motion of pendulum cannot be regarded as a simple harmonic motion. Here we use the electrostatic feedback control to solve this problem, as shown in Fig. 15.1. When the release tips are released, there is a net adhesion force F_r between the release tips and the test mass due to the asymmetric releasing and adhesion behavior. Then, the bias voltage V_b is applied to the capacitor plate that is installed on the top of the compound pendulum, and the control voltage $\pm V_f$ is applied to the nearby capacitor plates to generate a feedback electrostatic force. Under this condition, the pendulum remains at the initial equilibrium position by the electrostatic force F_v, which is equal to the adhesion force F_r and the restoring force of the pendulum, and it can be expressed as:

$$F_v = F_r + \frac{K\Delta\theta}{b}, \tag{15.2}$$

where $\Delta\theta$ is the residual angle of the pendulum with respect to the initial equilibrium position. The electrostatic force generated by the differential capacitance sensor can be expressed as:

$$F_v = \frac{2C}{d} V_b V_f, \tag{15.3}$$

where C is the capacitance facing the plate and d is the space between the plates at the equilibrium position, respectively. According to the control voltage V_f applied to the capacitor plate, the adhesion force to be measured can be written as:

$$F_r = \frac{2C}{d} V_b V_f - \frac{K\Delta\theta}{b}. \tag{15.4}$$

Table 15.1 The error budget of release impulse based on the compound pendulum

Error term	Design value	Accuracy requirement	Contribution error (kg·m/s)
Capacitance C	3.560 pF	0.001 pF	2.8×10^{-9}
Space d	4.00 mm	0.01 mm	2.5×10^{-8}
Bias voltage V_b	70.00 V	0.01 V	1.4×10^{-9}
Control voltage V_f	90.00 V	0.01 V	1.1×10^{-9}
Stiffness K	0.3000 Nm/rad	0.0005 Nm/rad	1.7×10^{-8}
Angle $\Delta\theta$	9.70 μrad	0.01 μrad	1.0×10^{-8}
Force arm b	80.00 mm	0.02 mm	2.5×10^{-9}
Duration time t	1.00 ms	0.01 ms	1.0×10^{-7}
Impulse I	1×10^{-5} kg·m/s	–	1.1×10^{-7}

15.3 Feasibility Analysis

Based on the introduction of the basic principle and the design of the measurement, the error budget is shown in Table 15.1. The designed release impulse of the test mass is 1×10^{-5} kg·m/s, with an uncertainty of 1.1×10^{-7} kg·m/s. The largest uncertainty is 1×10^{-7} kg·m/s that is contributed by the duration time of the adhesion force. This scheme provides an approach to simulate the on-orbit release process and also to study the adhesion force between the release tip and the surface of the test mass.

As can be seen from the table, the largest source of error that limits the measurement sensitivity is the duration time. Due to the existence of complex physical field in the experiment, especially the influence of ground vibration, there are inevitable errors in the experiment. We will conduct environmental modulation experiments later to improve the reliability of the measurement system.

15.4 Conclusion

The compound pendulum with the electrostatic feedback control is proposed to measure the release impulse in TianQin project. The pendulum has only a swing mode, whose attitude can be adjusted by the counterweights. And the stiffness of the compound pendulum can be calibrated with the gravity calibration method. The system has good stability because it is at the lowest point of gravitational potential energy. When the designed release impulse of the test mass is 1×10^{-5} kg·m/s, the expected measurement uncertainty is 1.1×10^{-7} kg·m/s. Under the electrostatic feedback control, the compound pendulum dose not move, and the adhesion force can be measured from the electrostatic force. This reduces the nonlinear effect of the pendulum. The measurement device will be built and the testing will be performed in future, and we will test the long-term stability and reliability of the system.

References

1. J. Luo et al., TianQin: ASpace-borne gravitational wave detector. Class. Quant. Grav. **33**, 030015 (2016)
2. J. Luo et al., The first round result from the TianQin-1 satellite. Class. Quant. Grav. **37**, 185013 (2020)
3. Mei, J, W., et al. The TianQin project: Current progress on science and technology. Prog. Theor. Exp. Phys. 05A107 (2021)
4. A. Zambotti, *Ground Testing and In-Flight Performance of a Space Mechanism* (University of Trento, Italy. Ph.D.thesis, 2019)
5. D. Bortoluzzi et al., Investigation of the in-flight anomalies of the LISA Pathfinder Test Mass release mechanism. Adv. Space Res. **68**, 2600–2615 (2021)
6. C. Zanoni et al., Summary of the results of the LISA-Pathfinder Test Mass release. J. Phys. Conf. Ser. **610**, 012022 (2015)
7. C. Zanoni et al., Experimental-analytical qualification of a piezoelectric mechanism for a critical space application. IEEE/ASME Trans. Mech. **20**, 427–437 (2015)
8. A. Lou et al., Theoretical calculations and experimental measurements on the two-component Au-Pt alloys with ultralow magnetic susceptibility. Phys. Rev. Appl. **19**, 034080 (2023)
9. Zou, Y, T., et al. Ground testing of release impulse for the aluminum cubic test mass with a compound pendulum for the TianQin project. Rev. Sci. Instrum. **94**, 094506 (2023),
10. H. Xu et al., A compound pendulum for thrust measurement of micro-Newton thruster. Rev. Sci. Instrum. **93**, 064501 (2022)

Chapter 16
The Measurement of Release Impulse and Impulse Moment with Different Types of Pendulums on the Ground

Yongqi Liu, Lingling Yang, Wencan Li, Xin Li, and Qing Li

Abstract In TianQin project, the release impulse of the test mass should be no more than 1×10^{-5} kg m/s and the release impulse moment should be no more than 5×10^{-8} kg m^2/s, respectively. In order to precisely measure the release impulse and impulse moment on the ground, we propose three kinds of pendulum systems to simulate the caging and release process of the test mass, they are the compound pendulum, the single pendulum, and the two-stage pendulum. Through mathematical modeling and dynamics analysis of these systems, we obtain the motion modes of the pendulum and analyze the feasibility of release impulse and impulse moment measurement.

16.1 Introduction

The TianQin project is one of the space-borne gravitational wave detection programs [1, 2]. In this project, the test mass will be carried by a specially designed mechanism, which sends the test mass into the orbit and completes the entire caging and releasing process. After the test mass is released, it is controlled by the electrostatic force and then goes into the free fall [3]. Since the electrostatic force is very small, the release impulse of the test mass should be no more than 1×10^{-5} kg m/s and the release impulse moment should be no more than 5×10^{-8} kg m^2/s, respectively.

The releasing process in space contains three stages [4]. In ground testing of release impulse and impulse moment, we use the pendulums to simulate the in-orbit situation. Compared with other methods, such as the air flotation and the electromagnetic suspension, the pendulum introduces a smaller perturbation, and is more stable and controllable. When the test mass is suspended by a thin fiber, the pull of the fiber counteracts the gravity of the test mass. In this work, we discuss three kinds of

Y. Liu · L. Yang · W. Li · X. Li · Q. Li (✉)
MOE Key Laboratory of Fundamental Physical Quantities Measurement & Hubei Key Laboratory of Gravitation and Quantum Physics, PGMF and School of Physics, Huazhong University of Science and Technology, Wuhan, China
e-mail: qingli@hust.edu.cn

J. Luo et al. (eds.), *Proceedings of the 7th International Workshop on the TianQin Science Mission; 25–26 April 2024, Hong Kong SAR, China*, Springer Proceedings in Physics 428, https://doi.org/10.1007/978-981-96-7803-7_16

pendulum systems which are often used in measurement of the release impulse and impulse moment: the compound pendulum, the single pendulum, and the two-stage pendulum in turn.

16.2 Modeling and Mathematical Derivation

16.2.1 Compound Pendulum

As shown in Fig. 16.1, the test mass is installed on the frame of the compound pendulum, which is symmetrically suspended by two strips [5–7]. Each strip has a length of 1 mm, a width of 8 mm, and a thickness of 20 μm. The test mass is locked and released by two tips. After released, the pendulum swings with an angle of θ along the release direction, while the rotation around the vertical axis can be ignored. This design improves the measurement precision of the release impulse along the release direction effectively.

Under the action of release impulse P, the compound pendulum acquires an initial angular velocity $\dot{\theta}_0$ and an initial angle, then we can obtain:

$$Pl = I\dot{\theta}_0 \tag{16.1}$$

where l is the force arm of the release tip and I is the moment of inertia of the compound pendulum, respectively. After released, the motion equation of the compound pendulum is:

$$I\ddot{\theta} + \lambda\dot{\theta} + K\theta = 0 \tag{16.2}$$

where $\lambda = 2I\zeta$ is the damping factor, ζ is the damping coefficient, K is the stiffness of the compound pendulum, respectively. The solution is:

$$\theta = \theta_r e^{-\zeta t} \cos(\omega t + \varphi) \tag{16.3}$$

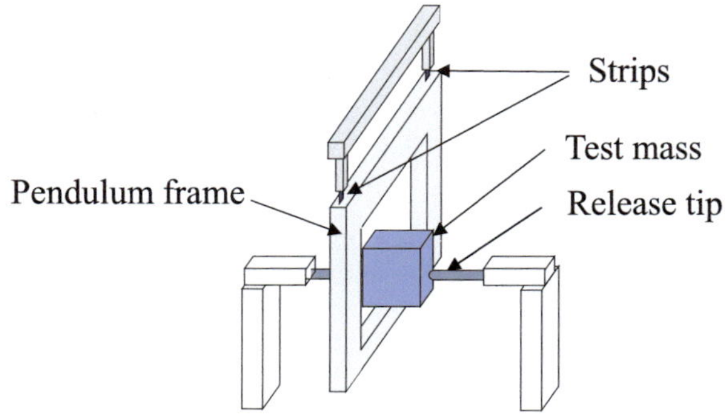

Fig. 16.1 Diagram of compound pendulum. The test mass is installed on the frame of the pendulum, which is symmetrically suspended by two strips. The test mass is locked and released by two tips. After released, the pendulum swings with an angle of θ

where θ_r is the initial amplitude of the compound pendulum, ω is the oscillation frequency, $\omega^2 = \omega_0{}^2$-ζ^2, ω_0 is the natural frequency without damping, $\omega_0{}^2 = K/I$, φ is the initial phase of the compound pendulum, respectively.

The release impulse acting on the compound pendulum can be expressed as:

$$P = \frac{K\theta_r}{l(\omega^2 + \zeta^2)\sqrt{\left(\frac{1}{\omega\tan\varphi+\zeta}\right)^2 + \left(\frac{1}{\omega} - \frac{\zeta}{\omega^2\tan\varphi+\zeta\omega}\right)^2}} \tag{16.4}$$

The typical parameters selected from the completed experiments are as follows d[6]: the torsion spring constant K is 0.34 Nm/rad, the initial amplitude θ_r is 1×10^{-4} rad, the force arm l is 8×10^{-2} m, the damping coefficient ζ is 1×10^{-3}/s, the frequency of pendulum ω is 9.5 rad/s, the initial phase of pendulum φ is 1.6 rad. According to the work [6], we estimate that the measurement uncertainty of the release impulse is on the order of magnitude of 10^{-7} kg m/s. Moreover, if the experiment is conducted in a high vacuum environment, the uncertainty can be further decreased.

16.2.2 Single Pendulum

As shown in Fig. 16.2, the single pendulum uses a thin fiber to suspend the test mass. The test mass is locked and released by the opposite tips. Compared with the compound pendulum, the single pendulum has three directions of motion, which includes two translations along the horizontal directions and the rotation around the vertical axis. The translations of the single pendulum can be used to measure the release impulse, while the latter is used to study the release impulse moment.

Similar to the compound pendulum, the release impulse of the test mass can be measured by using the single pendulum. The parameters for the release impulse measurement are shown in Table 16.1, These parameters come from the experimental

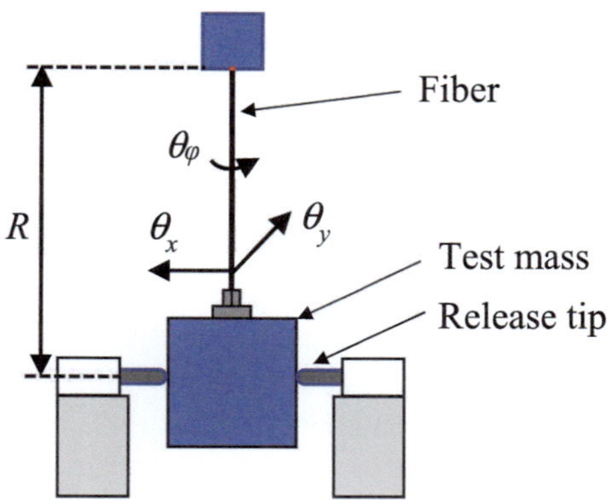

Fig. 16.2 Diagram of single pendulum. The test mass is suspended by a thin fiber, it is locked and released by two tips. After released, the single pendulum has three directions of motion, which includes two translations along the horizontal directions and the rotation around the vertical axis

Table 16.1 The error budget of impulse measurement. The last column is in units of kg m/s

Parameter	Design value	Design error	Uncertainty of impulse
Moment of inertia J	6.9×10^{-1} kg m^2	6.9×10^{-4} kg m^2	0.1×10^{-7}
Force arm R	8.25×10^{-1} m	2×10^{-5} m	0.1×10^{-7}
Amplitude θ_r	3.5×10^{-6} rad	1×10^{-7} rad	2.8×10^{-7}
Damping efficient ζ	4.0×10^{-3}/s	4.0×10^{-6}/s	0.1×10^{-7}
Motion frequency ω	3.45/s	3.45×10^{-3}/s	0.1×10^{-7}
Initial phase φ	1.57 rad	1.57×10^{-3} rad	0.1×10^{-7}
Total			2.9×10^{-7}

Table 16.2 The error budget of impulse moment measurement. δL is in units of kg m^2/s

Parameter	Design value	Design error	δL
Moment of inertia J_φ	4.17×10^{-4} kg m^2	4.17×10^{-7} kg m^2	4.9×10^{-11}
Torsional amplitude φ_r	1.12×10^{-3} rad	1.25×10^{-3} rad	5.5×10^{-11}
Damping efficient ζ	4.0×10^{-3}/s	4.0×10^{-6}/s	0.1×10^{-11}
Motion frequency ω_φ	0.106/s	0.106×10^{-3}/s	4.9×10^{-11}
Initial phase ϕ	1.57 rad	1.57×10^{-3} rad	0.3×10^{-11}
Total			8.9×10^{-11}

plan that has completed the preliminary design. The expected uncertainty is 2.9×10^{-7} kg m/s.

Besides, the release impulse moment L can be expressed as:

$$L = \frac{J_\varphi \varphi_r}{\sqrt{\left(\frac{1}{\omega_\varphi \tan\phi + \zeta}\right)^2 + \left(\frac{1}{\omega_\varphi} - \frac{\zeta}{\omega_\varphi^2 \tan\phi + \zeta\omega_\varphi}\right)^2}} \tag{16.5}$$

where J_φ is the moment of inertia of the single pendulum around the vertical axis, φ_r is the initial amplitude of the torsion pendulum, ζ is the damping coefficient, ω_φ is the torsion frequency, ϕ is the initial phase, respectively. The corresponding parameters is shown in Table 16.2, and the expected uncertainty is 8.9×10^{-11} kg m^2/s.

16.2.3 Comparative Analysis Section

The above two methods respectively use the compound pendulum and the single pendulum to measure the release impulse of the test mass. When measuring with a compound pendulum, the test mass only swings in one direction, this makes the measuring system simpler and we can get more accurate measurements in a single degree of freedom with less coupling interference. On the flip side, motion in a single degree of freedom means that the effects of locking and releasing in other directions cannot be studied. This differs from the situation in orbit.

The single pendulum is relatively simple in structure and it has more degrees of freedom. A test mass suspended by a single pendulum can move freely in two orthogonal directions and twist around the suspension wire. This experimental scheme allows the measurement of motion on more degrees of freedom without introducing more external disturbances. However, simultaneous motion in multiple degrees of freedom will introduce coupling between measurements of different motions. The above two experimental plans each have their pros and cons. Overall, the single pendulum plan better meets the requirements of the TianQin Project for ground testing because the single pendulum system has more degrees of freedom and is more in line with actual on-orbit conditions..

16.3 Summary

In this paper, two types of pendulums with different configurations are introduced, and the motion modes of the test mass in these systems are analyzed. Currently, the use of a compound pendulum to measure the release impulse of the test mass has achieved certain results. In ground tests, the release impulse meets the requirements of the TianQin project, with measurement accuracy reaching the level of 10^{-7}. The measurement experiment using a single pendulum is currently in progress.

References

1. J. Luo et al., TianQin: a space-borne gravitational wave detector. Cl. Quantum Grav. **33**, 035010 (2016)
2. J. Luo et al., The first round result from the TianQin-1 satellite. Cl. Quantum Grav. **37**, 185013 (2020)
3. C. Zanoni et al., Summary of the results of the LISA-pathfinder test mass release. J. Phy.: Conf. Ser. **610**, 012–022 (2015)
4. D. Bortoluzzi et al., Prediction of the LISA-Pathfinder release mechanism in-flight performance. Adv. Space Res. **51**, 1145–1156 (2013)
5. H. Xu et al., A newly designed decoupling method for micro-Newton thrust measurement. Rev. Sci. Instrum. **94**, 014504 (2023)
6. Y.T. Zou et al., Ground testing of release impulse for the aluminum cubic test mass with a compound pendulum for the TianQin project. Rev. Sci. Instrum. **94**, 094506 (2023)
7. H. Xu et al., A compound pendulum for thrust measurement of micro-Newton thruster. Rev. Sci. Instrum. **93**, 064501 (2022)

Chapter 17
The GPRM Working Procedure Design and Verification for Space-Based Gravitational Wave Observatory

Qian Wen, Kian Hong Low, and Zhaokui Wang

Abstract In the context of space missions, extensive ground testing is conducted to ensure functionality and performance before launch. The gravitational reference sensor, crucial for the quality of detected gravitational wave signals, relies on a critical payload known as the test mass. Meanwhile, the grabbing, positioning, and release mechanism manages the injection of the test mass from a caged configuration into a free-falling state. It is essential to characterize the dynamic behaviors of the test mass during release, with stringent constraints on residual momentum. Seven working modes—test, launch, release, calibration, scientific, recapture, and standby mode—are proposed by this paper, for the aforementioned mechanism. These modes complement each other to address all potential situations and emergencies in space. Building on these seven modes, this paper also introduces a hardware-in-the-loop simulation method, combining physical mechanisms and algorithms to address the congruity between ground and space conditions. Evaluation methods are presented to optimize and iterate the parameters for each working mode.

17.1 Introduction

Gravitational wave (GW) observation provides an extraordinary and unique perspective for scientists to uncover the fundamental laws of the universe, study the history and characteristics of black holes, and verify the general theory of relativity. However, restricted by several limitations, ground-based GW observatories are unable to detect GW which frequency is lower than 10 Hz, whereas frequency bands from 0.1 Hz to 0.1 mHz are consisted with stronger and abundant GW sources. In order to tackle the constraints mentioned above, several space-borne projects, such as, LISA [1], TianQin [2, 3], Taiji [4], and DECIGO [5] have thus been proposed or even under development.

Q. Wen · K. H. Low · Z. Wang (✉)
School of Aerospace Engineering, Tsinghua University, Beijing, China
e-mail: wangzk@tsinghua.edu.cn

J. Luo et al. (eds.), *Proceedings of the 7th International Workshop on the TianQin Science Mission; 25–26 April 2024, Hong Kong SAR, China*, Springer Proceedings in Physics 428, https://doi.org/10.1007/978-981-96-7803-7_17

For a space-based GW observatory, the key to successfully detecting GW signals in space is the implementation of a laser interferometer, which requires reference points provided by purely free-flying test masses (TMs). Here, the gravitational reference sensor (GRS) serves as the core component responsible for measuring and controlling, while also shielding the TMs from external force noise [6]. Among the components of the GRS, the Grabbing Positioning and Release Mechanism (GPRM) plays a pivotal role in precisely constraining and adjusting the position of the TM throughout the entire mission duration [7]. It ensures the successful transition of the TMs from the locking phase into the Drag-Free phase.

The LISA Pathfinder (LPF), which served as a demonstration mission prior to LISA, integrated two GRSs, each hosting a cubic-shaped TM. One of the critical tasks for LPF was to successfully inject these two TMs into free-flying trajectories. Extensive experiments were conducted to understand the dynamics and limits of the GPRM for future injections on LISA [8, 9]. However, to our knowledge, an overall optimal release strategy has not been found, as the two GPRMs on LPF behaved differently, and the TMs were only captured after impacts. Fortunately, the LISA team gathered numerous valuable insights from these experiments, enabling them to enhance their future mission [10, 11]. This again emphasized the importance of developing an on-ground verification workflow to ensure the successful implementation of TM release in orbit.

The key to on-ground verification is to measure the momentum transferred to the TM. The University of Trento [12] has thus developed a Transferred Momentum Measurement Facility (TMMF), where the TM is suspended by a metal fiber to counteract the effects of gravity. Similarly, TianQin's team has performed a series of measurements using torsion and compound pendulums [13, 14]. Efforts have included measuring release impulse, assessing remanent magnetic properties, and developing a charge management scheme, all aimed at providing a more comprehensive understanding to better support future locking and release efforts by the GPRM in orbit [15, 16].

Based on the aforementioned efforts and results, this paper aims to design a complete workflow for the GPRM, which we have decomposed into seven working modes. These modes complement each other, aiming to ensure the high reliability and sustainability of the GPRM in orbit. External disturbances, including adhesion forces, collisions, deformation, and asymmetrical release, are carefully considered in this design. An on-ground verification scheme has also been proposed to address the congruity theory between ground and space conditions. We hope our work provides insightful information for the future implementation of the GPRM in space-based GW observatories.

17.2 Principles of Mechanism Operation

The GPRM is composed of three substructures: the caging mechanisms, the plungers, and the release tips. Each substructure operates within a specific pre-load force range to maintain the central position of the TM within the GRS. As shown in Fig. 17.1, there are eight locking claws (LoCs), each corresponding to one vertex of the cubic TM. During the phases of rocket launching and orbit transferring, the TM is securely locked by these eight LoCs with a pre-load force around 1500N, depending on the resonance modes of the selected rocket in the actual mission.

After the satellite has entered its intended orbit, the TM needs to be released gently for the implementation of scientific detection. This requires ensuring that its linear and angular velocities remain within a given threshold. To facilitate this, the LoC will hand over the TM to the plungers, reducing the adhesion force and ensuring a symmetrical release process. The pre-load forces introduced by the plungers, controlled by piezo-walk actuators, are usually three orders of magnitude lower than those of the LoC, docking with the TM on an intended notch.

Inside the plungers, there are axially movable release tips. Unlike the plungers, which have a large movement range to extend from outside into the electrode housing, the release tips have a smaller range, on the order of micrometers. The transition from plungers to release tips is a synchronous process, where the extension of the release tips is accompanied by the retraction of the plungers. The release tips are designed to produce a "point contact" with the TM, thus minimizing the contact area.

The release or relock process of the TM by the GPRM is complex, involving the synchronous and symmetrical movement of each substructure and the retracting speed. It also needs to account for external disturbances such as adhesion forces, electromagnetic forces, and thermal forces. Whether collisions between the GPRM's substructures and the TM can be utilized to enhance the release process to meet the highly demanding release standards is also a consideration.

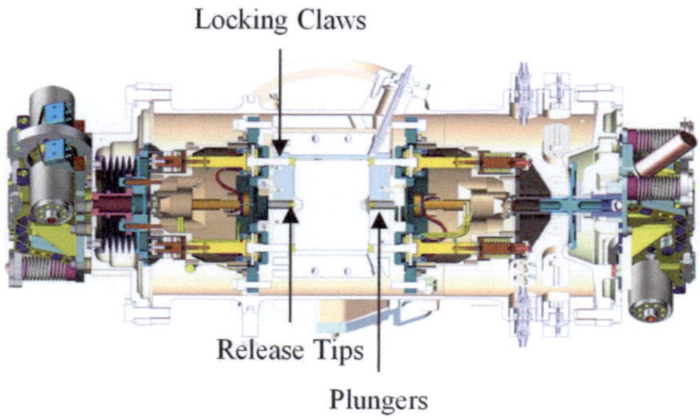

Fig. 17.1 The diagram of GPRM [17]

17.3 Design and Verification of Working Modes

17.3.1 Types of Working Modes

Based on the operational characteristics and mission requirements, we have developed seven distinct working modes to ensure the GPRM is highly reliable and repeatable. These modes are defined as follows:

Test mode. The test mode, also known as the pre-launch mode, is essential for verifying the functionality of the GPRM. This mode involves ground-based verification, simulating the microgravity of the space environment, with particular emphasis on mechanisms such as adhesion force and deformation. Additionally, the impact of the on-orbit environment on the mechanism's performance must be comprehensively considered. Iterative optimization is achieved through multiple cycle testing processes, where each cycle provides data for adjustments and improvements. Extensive ground-based tests and simulations ensure the GPRM performs reliably under expected space conditions, identifying and mitigating potential issues to ensure it meets operational requirements once deployed in space (Fig. 17.2).

Launch mode. This mode aims to ensure the TM's integrity during the strong vibrations introduced by rocket launching, including its symmetry, surface smoothness, and coating integrity, which are crucial for providing a stable reference point for the interferometer during GW detection. An optimal locking force provided by the clamping mechanism has been predefined on the ground, carefully calibrated and optimized to prevent the TM from being collided or deformed. According to the design and test requirements, the LoCs generate a locking force of approximately 1500N, while the plungers and release tips contribute locking forces of 60N and 0.6N, respectively. This approach ensures the TM remains securely docked, maintaining its integrity and providing a stable reference point for accurate GW signal detection.

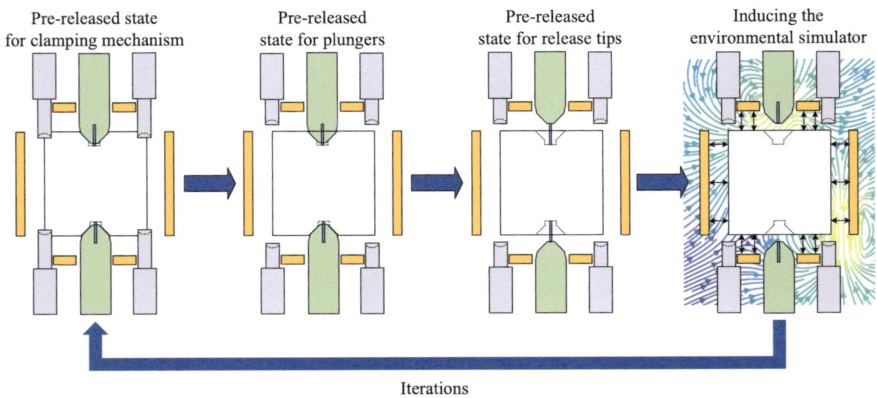

Fig. 17.2 The schematic process for test mode

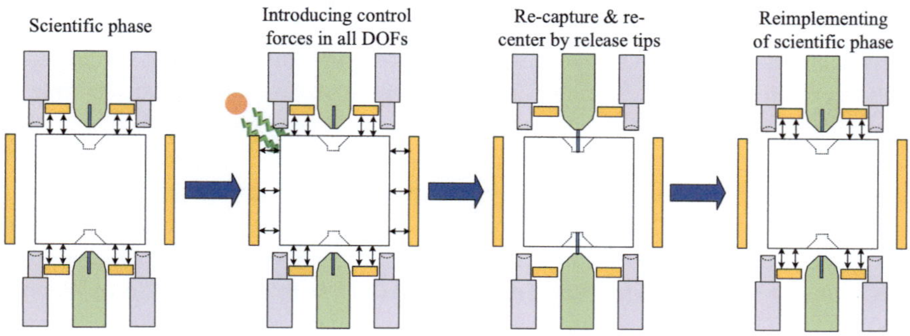

Fig. 17.3 Translation motion of the TM driven by the plungers to calibrate

Calibration mode. This mode will be conducted both before and after the scientific phase for the purpose of maintenance. By recalibrating, most drifts or deviations that may have occurred could be correct, preserving the integrity of the scientific data. During calibration, only the plungers will be involved due to its larger displacement range. The plunger will dock with the TM and move translationally along the non-sensitive axes, allowing the capacitance sensor to recalibrate by measuring the TM's displacement inside the electrode housing (EH). The sensor measures the displacement and torque variations of the TM continuously, and using the recorded data to verify constructed model by the theoretical plunger's translation distance and the sensor's measurement deviation. The system algorithm then adjusts the sensor's coefficients by calculating ideal and measured values, and the updated coefficients are applied to ensure the accuracy of scientific mode. The key to successful calibration is the precise control of the plunger's movement, ensuring the TM's displacement matches the signals detected by the capacitance sensor accurately (Fig. 17.3).

Release mode The release mode is activated once the satellite has entered its intended orbit and is ready to detect GW signals. This mode involves sequencing the operations of the substructures of the GPRM. The goal is to minimize residual linear and angular momentum to accommodate the tiny gaps between the TM and the EH. During the release process, the clamping mechanism must gently disengage from the TM without imparting significant forces that could alter its position. Simultaneously, the plunger must be extended in a syn-chronized manner to capture the TM. To further reduce contact surfaces between the GPRM and the TM, the release tips will act as the final releasing mechanism, taking over once the plunger has disengaged. Thorough testing of each component and sequence ensures a seamless transition from the pre-launch phase to the scientific phase, with the TM perfectly released and within the controllable threshold provided by the electrode.

Scientific mode In space-based GW detection missions, the scientific mode is the core operational state dedicated to measuring extremely weak GW signals. This mode is initiated only under stringent conditions: (1) the TM must be precisely centered within the EH, satisfying strict position and attitude constraints; (2) electrostatic control forces and torques acting on the TM must be in equilibrium; and (3) all sensors and control subsystems must be operating nominally. Owing to the ultra-high sensitivity of the space-based laser interferometer, the system is susceptible to external disturbances such as solar storms and micro-meteoroid impacts. To maintain mission continuity and ensure measurement precision, rapid mode transitions and protective measures must be promptly executed upon detection of any anomalies.

Recapture mode The recapture mode is typically triggered under the conditions such as the completion of an operational cycle, unexpected disturbances during operations, or the need for recapture due to excessive release impulse or angular momentum. The primary objective of this mode is to ensure that the TM is safely and stably returned to a locked state in all scenarios. This mode is also typically designed to handle extreme cases, such as when a collision occurs between the EH and TM. In situations where the EH loses control of the TM, a new strategy must be implemented to recapture the TM and prevent the situation from deteriorating further. In this case, we utilize the translational motion of the release tips to rotate the TM into a centralized attitude. Once the gaps between the TM and EH are within the control threshold, we then introduce full DOFs electrostatic force, followed by engagement performed by the plungers. This mode requires a highly realistic dynamics model of the TM and the GPRM to implement precise control for re-capturing the TM.(Fig. 17.4).

Standby mode. For TianQin, a strategic operational cycle termed '3 months on + 3 months off' has been proposed by Prof. Luo and his team[18] to avoid the satellite's optical telescope being directly exposed to sunlight, resulting in the detector not operational for half of the year. At the end of each scientific phase, the TM must be re-captured to perform essential maintenance processes, such as calibration mode. Thus, we introduced this recapture mode, which is critical for maintaining the system's high precision and reliability. Unlike the previous scientific mode, which involves temporary engagement, this mode provides a long-lasting docking action of the TM at its central position. In addition to introducing electrostatic forces in all DOFs of the TM, we extend the plungers to securely capture and lock the TM, rather than relying solely on the release tips. This comprehensive approach ensures the TM is

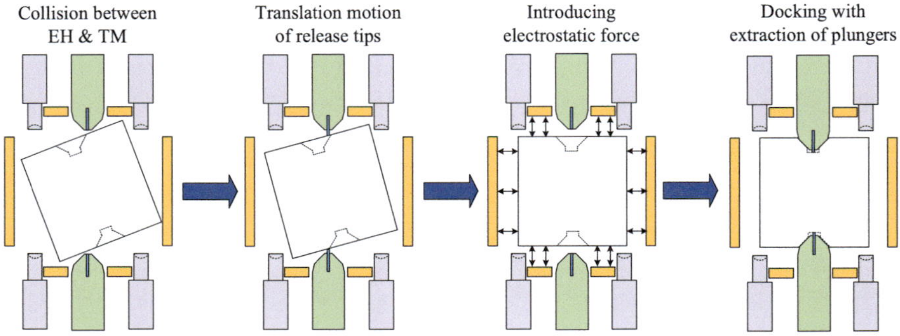

Fig. 17.4 Re-capture the collided TM by extraction of release tips

firmly held, allowing thorough maintenance and recalibration, thereby sustaining the GRS system's performance and accuracy over the long term.

17.3.2 Methodology and Verification of HIL Simulation

Based on the seven working modes outlined, it is essential to construct a verification platform for these modes. Given the challenge of accurately replicating the actual in-space environment encountered by the TM, we propose a Hardware-in-the-Loop (HIL) simulation method to ensure successful implementation. Specifically, the orbital environment is numerically modeled to calculate the disturbances the TM will experience during different phases. These disturbances are generated by algorithms to simulate the "actual" space environment, as full replication of these conditions on the ground is impossible. The phases are categorized by disturbance magnitude into the Transfer and Injection Phase (TrIP), Scientific Operations Phase (ScOPe), and Standby and Maintenance Phase (StaMP) (Fig. 17.5).

For the HIL simulation, the hardware component consists of the GPRM, a real model that performs all the motions of its substructures according to each working mode. Concurrently, we will develop a digital twin system that combines the dynamical and kinematic models of the GPRM and TM with the environmental simulation algorithm. This approach allows synchronous comparison and evaluation of experimental results with those from the full digital model, facilitating iterative refinement of the working modes' parameters. Additionally, we propose an evaluation model considering key factors such as stability, response speed, robustness, and accuracy to assess the HIL simulation results comprehensively.

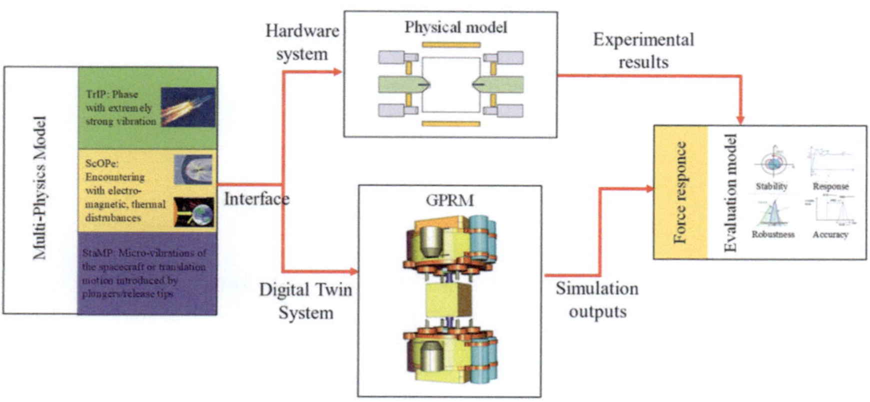

Fig. 17.5 The Hardware-in-the-loop simulation workflow

17.4 Conclusion and Future Work

In this paper, we propose seven working modes for the locking, releasing, and capturing actions of the GPRM. These modes are thoroughly discussed to address all nominal situations and potential emergencies in space. Additionally, to validate these modes on the ground before launch, we introduce a HIL simulation workflow and an evaluation method to bridge the gap between ground and space conditions. Our team is also developing a multi-physical space environment system, incorporating factors such as ambient magnetic fields, space plasma penetration, thermal fluctuations, and mechanical vibrations. This system aims to provide a more realistic representation of the on-orbit environment, enhancing the accuracy and reliability of the simulation data.

However, several challenges or technical limitations need to be addressed before fully implementing this system, such as:

1. Mechanical Vibrations: Mechanical vibrations varied throughout the mission, from launch to orbital transfer and attitude control. A vibration model is needed to assess the GPRM's functionality under both high- and low-frequency vibrations.
2. Electromagnetic Field Data: Space engineering seldom covers far-Earth or Sun-Earth environments, lacking sufficient electromagnetic field data, making precise inversion difficult, especially for low-noise tasks like space-based gravitational wave detection.
3. Environmental Simulation: Accurately replicating and applying multi-physical space environments to the locking and release mechanism during ground-based semi-physical simulations is a significant technical limitation.

Our future work will focus on addressing these limitations and further examining the proposed modes using the HIL simulation framework.

Acknowledgements We acknowledge the support from the National Key R&D Program of China (Grant numbers: 2023YFC2205601 and 2022YFC2204204).

References

1. P. Amaro-Seoane, J. Andrews, M.A. Sedda et al., Astrophysics with the laser interferometer space antenna. Living Rev. Relativ. **26** (2023)
2. J.W. Mei, Y.Z. Bai, J.H. Bao et al., The Tianqin project: Current progress on science and technology. Prog. Theor. Exp. Phys. (2021)
3. J. Luo, L.S. Chen, H.Z. Duan, Tianqin: a space-borne gravitational wave detector. Cl. Quantum Grav. **33** (2016)
4. W.H. Ruan, Z.K. Guo, R.G. Cai et al., Taiji program: gravitational-wave sources. Int. J. Mod. Phys. A **35** (2020)
5. S. Kawamura, M. Ando, N. Seto et al., Current status of space gravitational wave antenna DECIGO and B-DECIGO. Prog. Theor. Exp. Phys. (2021)

6. S. Vidano, C. Novaro, M. Pagone et al., The LISA DFACS: model predictive control design for the test mass release phase. Acta Astronaut. **193**, 731–743 (2022)
7. D. Bortoluzzi, P.A. Mausli, R. Antonello et al., Modeling and identification of an electro-mechanical system: the LISA grabbing positioning and release mechanism case. Adv. Space Res. **47**(3), 453–465 (2011)
8. D. Bortoluzzi, D. Vignotto, A. Zambotti et al., In-flight testing of the injection of the LISA Pathfinder test mass into a geodesic. Adv. Space Res. **67**(1), 504–520 (2021)
9. D. Bortoluzzi, D. Vignotto, E. Dalla Ricca et al., Investigation of the in-flight anomalies of the LISA Pathfinder test mass release mechanism. Adv. Space Res. **68**(6), 2600–2615 (2021)
10. M. Tomasi, E. Dalla Ricca, D. Vignotto et al., Development of a dynamical model of a release mechanism for in-flight performance prediction. Adv. Space Res. **72**(11), 4950–4964 (2023)
11. D. Vignotto, D. Bortoluzzi, Impact dynamics of a free-falling reference test mass in space. Int. J. Impact Eng. 186 (2024)
12. D. Bortoluzzi, M. Benedetti, L. Baglivo et al., Measurement of momentum transfer due to adhesive forces: on-ground testing of in-space body injection into geodesic motion. Rev. Sci. Instrum. **82**(12), 125107 (2011)
13. H. Xu, Y. Gao, Q.B. Mao et al., A compound pendulum for thrust measurement of micro-Newton thrusters. Rev. Sci. Instrum. **93**(6), 064501 (2022)
14. Y.T. Zou, W.C. Li, Q.B. Mao et al., Ground testing of release impulse for the aluminum cubic test mass with a compound pendulum for the TianQin project. Rev. Sci. Instrum. **94**(9), 094506 (2023)
15. F.C. Yang, Y.Z. Bai, W. Hong et al., Investigation of charge management using UV LED device with a torsion pendulum for TianQin. Cl. Quantum Grav. **37**(11), 115005 (2020)
16. J.H. Xu, Q. Liu, X. Luo et al., Measuring ac magnetic susceptibility at low frequencies with a torsion pendulum for gravitational-wave detection. Phys. Rev. Appl. **18**(4), 044010 (2022)
17. B. Zahnd, M. Zimmermann, R. Spörri, LISA-Pathfinder cage and vent mechanism—development and qualification. in *Proceedings of the 15th European Space Mechanism and Tribology Symposium (ESMATS)* (2013), pp. 1–7
18. H.Y. Chen, C. Ling, X.F. Zhang et al., Thermal environment analysis for TianQin. Cl. Quantum Grav. 38 (2021)

Chapter 18
Thrust Noise Rejection of MDIT and MDCFT by Feedback Control System Faced with Space Gravitational Waves Detection

Xiang Niu, Hongbo Zhou, Hui Liu, and Daren yu

Abstract Drag free control system, as a key part of space gravitational waves detection satellite, needs to offset non-conservative forces on the satellite and reduce its residual acceleration. As the actuator of drag-free control system, the micro propulsion system needs to output super low-noise thrust to realize the detection goal. The microwave discharge ion thruster (MDIT) and microwave discharge cusped field thruster (MDCFT) are two minimized electric thrusters designed by Harbin Institute of Technology (HIT), which both can realize 1-100μN thrust range. However, high thrust noises in the low frequency band limit their applications on the space GWs detection task. Analyses of thrust noise sources show the main disturbances are ionization processes. Based on it, a feedback control system is designed according to the characteristics of the two thrusters. Then an experimental platform is established on ground to validate that the thrust noises of MDIT and MDCFT both can be reduced below $0.1\mu N/Hz^{1/2}$ in 1mHz to 1Hz after adopting the feedback control system.

18.1 Introduction

Space gravitational waves (GWs) detection is a kind of high-precision scientific experiment aiming to detect the low-frequency GWs by measuring the optical length variations between two proof masses [1]. There are many space detection plans including Laser Interferometer Space Antenna (Lisa) [2], DECi-hertz Interferometer Gravitational wave Observatory (DECIGO) [3], Taiji plan [4] and TianQin plan [5]. The TianQin plan relies on three identical satellites on geocentric orbits to form a triangular configuration, whose semi-major axis is about 10^5km. In the detection plan, drag-free control system is a key part, which consists of displacement sensor,

X. Niu · H. Zhou · H. Liu (✉) · D. yu
Harbin Institute of Technology, Harbin, China
e-mail: huiliu@hit.edu.cn

J. Luo et al. (eds.), *Proceedings of the 7th International Workshop on the TianQin Science Mission; 25–26 April 2024, Hong Kong SAR, China*, Springer Proceedings in Physics 428, https://doi.org/10.1007/978-981-96-7803-7_18

drag-free controller and micro thruster [6]. There is a freely floating test mass in the drag-free satellite. The displacement sensor can detect the relative displacement between the satellite and the test mass. The drag-free controller can generate proper thrust demand according to the relative displacement and make the micro propulsion system output low-noise thrust to offset the non-conservative forces on the satellite and reduce its residual acceleration.

As the key part of drag-free controller, the micro propulsion system should satisfy many requirements, including wide thrust rang (5 ~ 100μN), low thrust noise (0.1μN/$Hz^{1/2}$), short response time (50 ms), high thrust resolution (0.1μN) and long lifetime (10000 h) [7]. There are some micro thrusters which have been launched to validate their feasibility. The LISA Pathfinder is launched by LISA choosing the colloidal thruster and cold gas thruster. However, a colloidal thruster cannot work due to obstruction of emitter, which reflects that its reliability is seriously weakened by its complex structure [8]. The cold gas thruster shows low specific impulse and slow response [9]. The Taiji-1 is launched by Taiji plan choosing radio-frequency ion thruster and Hall micro thruster, whose thrust noises are both higher than 0.1μN/$Hz^{1/2}$ [10, 11]. The TianQin-1 is launched by TianQin plan choosing cold gas thruster, which also faces similar problems like that in LISA Pathfinder [12]. Faced with space gravitational waves detection, the minimized microwave discharge ion thruster (MDIT) [13] and microwave discharge cusped field thruster (MDCFT) [14] are designed. Test results show that their ranges can both realize 1–100 μN, satisfying the requirements of space gravitational wave detection. However, the thrust noises of the two thrusters in 1 mHz ~ 1 Hz are both higher 0.1 μN/$Hz^{1/2}$. Test results show that high thrust noise is located in the low frequency band. There are many disturbances which can lead to high thrust noises like the fluctuations of flow rate, microwave power and DC voltage. It is hard to reduce the thrust noise by optimization of thruster. The feedback control is effective for low-frequency thrust noise rejection and can be used to solve the problem.

18.2 Characteristics of MDIT and MDCFT

18.2.1 Introduction of MDIT and MDCFT

The structure of MDIT and MDCFT are shown as Fig. 18.1. The MDIT utilizes the microwave to heat electrons based on electron cyclotron resonance, which can generate more high-energy electrons and enhance ionization process. The ions from ionization process can be accelerated by the strong electric field produced by grid system and produce thrust. Compared with traditional MDIT, the minimized MDIT designed by Harbin Institute of Technology (HIT) adopts smaller diameter and axial cusped magnetic field. The smaller diameter of discharge chamber contributes to lower thrust minimum. The axial cusped magnetic field can widen thrust range and increase its lifetime by reducing the interaction between plasma and wall. The grid

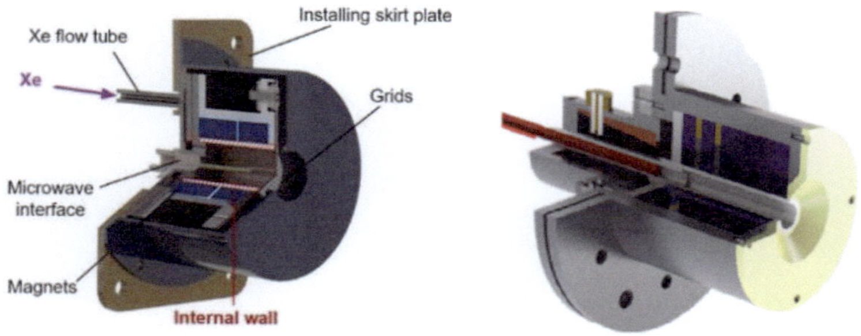

(a) MDIT (b) MDCFT

Fig. 18.1 Strcuture of MDIT and MDCFT

system can make the ion beam current measurable, which can improve the accuracy of thrust estimation.

The MDCFT utilizes the microwave to keep ignited during the minimum thrust operation by the coaxial transmission line resonator (CTRL). The CTRL can produce the strongest electric field at the open end of the cavity to ionize the neutral gas when at resonance. The MDCFT also adopts axial cusped magnetic field to increase its lifetime. Besides, a thermal emission cathode is adopted as the neutralizer for MDIT and MDCFT. This cathode works without propellant and can avoid the thrust deflection caused by propellant from cathode. Compared with the colloidal thrusters, the MDITs and MDCFTs are more reliable. Besides, their thrust range reach 1-100μN, which is wider than that of the colloidal thrusters. Compared with the cold gas thruster, the adjustment time of the microwave power to the thrust is on the order of milliseconds, which can improve the bandwidth of the feedback control system, enhance the noise rejection effect and thrust response velocity.

18.2.2 Thrust Noises of MDIT and MDCFT

We evaluate the thrust range by its electric parameters, which has been validated by high-precision thrust stand. Test results show that the MDIT [13] and MDCFT [15] can both realize 1-100μN range. The thrust noises of the two electric thrusters based on electric parameters under 100-μN operations are shown as Fig. 18.2. It can be seen that the thrust noises both increases with the decrease of frequency. The thrust noises of MDIT and MDCFT respectively reach $4\mu N/Hz^{1/2}$ and $6\mu N/Hz^{1/2}$, which are higher than the requirements of space gravitational waves detection.

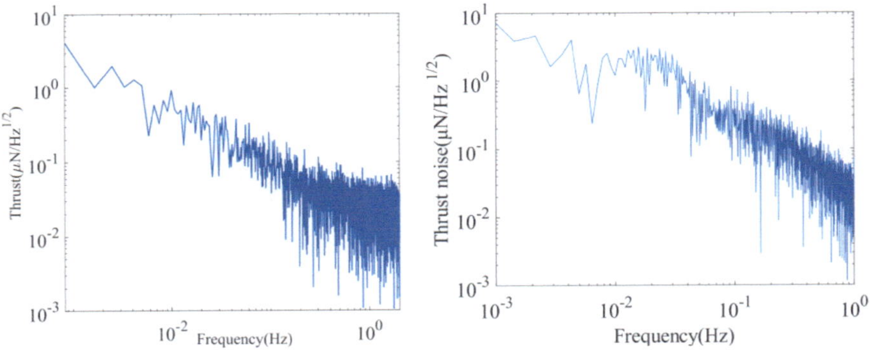

(a) MDIT (b) MDCFT

Fig. 18.2 Thrust noise under 100-μN operation

18.3 Design of Feedback Control System

There are two elementary processes for electric thrusters, namely ionization and acceleration process. The ion beam current and anode current can respectively reflect fluctuations of ionization processes for MDIT and MDCFT. The screen grid (SG) voltage and anode voltage can respectively reflect the fluctuations of acceleration processes for MDIT and MDCFT. Open-loop results show that noises of ion beam current and SG voltage are respectively 0.1 $mA/Hz^{1/2}$ and 1 $V/Hz^{1/2}$ at 1 mHz under 100-μN operation for MDIT. Open-loop results show that noises of anode current and anode voltage are respectively 0.26 $mA/Hz^{1/2}$ and 0.73 $V/Hz^{1/2}$ at 1 mHz under 100-μN operation for MDCFT. It can be evaluated from thrust formula that the thrust noises of MDIT caused by SG current and SG voltage are respectively 5.7 $\mu N/Hz^{1/2}$ and 0.05 $\mu N/Hz^{1/2}$. The thrust noises of MDCFT caused by anode current and anode voltage are respectively 7.5 $\mu N/Hz^{1/2}$ and 0.1 $\mu N/Hz^{1/2}$. The main noise sources of MDIT and MDCFT are both from their ionization processes [16]. Therefore, the ion beam current and anode current are chosen as feedback parameters for feedback control system.

We talk about the controlled variable for the feedback control system. There are three parameters for MDIT and MDCFT, namely flow rate, microwave power and voltage. The response time of flow rate is slow, which would limit the bandwidth of feedback controller. For MDIT, the influence of SG voltage on ion beam current is weak. The adjustment ability of anode voltage is strong for MDCFT. However, the anode voltage would be unstable when used to stabilize the anode current, which would seriously affect the stability of thrust from its acceleration process. The microwave power is an effective way to adjust ion beam current and anode current for the two electric thrusters and is proper as controlled variable to stabilize the ion beam current and anode current. The feedback control frame of MDIT and MDCFT are similar, shown as Fig. 18.3. It can be seen from slope of

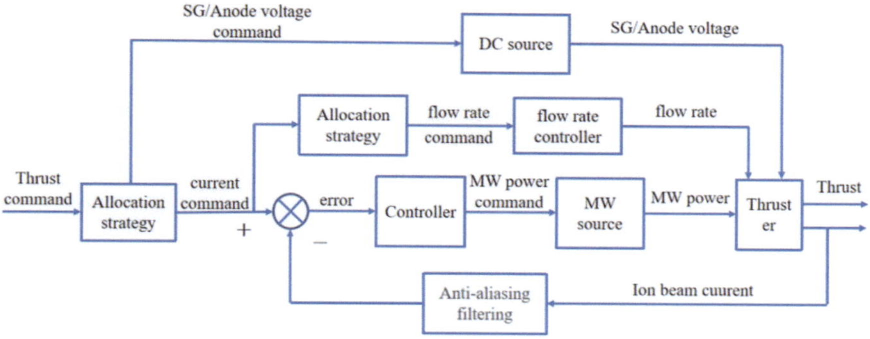

Fig. 18.3 Feedback control frame of MDIT

ion beam/anode current noise with frequency is 20dB/dec. Considering the bandwidth of the feedback controller is on the order of several decades of Hertz, the PI controller is enough to reduce the current noise to the requirements. The mechanism of the feedback control system to reduce the current noise is demonstrated here. We assume that the disturbances enhance the ionization process and cause the current higher than the set point. The error between the demand and the measurement result is negative, which can reduce the microwave power demand by feedback controller. The decrease of the microwave power demand can weaken the ionization process and reduce the current to its set point. Considering the control frequency is 1 kHz, the bandwidth of the feedback controller can ensure strong noise rejection effect in the low frequency band.

18.4 Experimental Results

We establish an experimental platform on ground to validate the effect of feedback control system. We utilize I-U transformer to transport ion beam current and anode current into corresponding voltage signal and then send the signal to single chip. The analogue to digital converter (ADC) of the single chip can transform the analogue to digital signal, which can be used to calculate command. The control code is written in single chip by C language. For the I-U transformer, 1mA current signal can be transformed into 1V voltage, whose accuracy is within 0.1%. To avoid the aliasing effect during the sampling process, an anti-aliasing filter is used to deal with feedback parameters. To avoid the negative influence of anti-aliasing filter on the response process of ion beam current and anode current, the cut-off frequency is set as 100Hz.

The adjustment ability of microwave power to ion beam current and anode current are divergent under different operations. Therefore, the parameters of PI controller are variable based on the stable and dynamic characteristics of model along the thrust adjustment path. The controller parameters are smaller for the operations where the

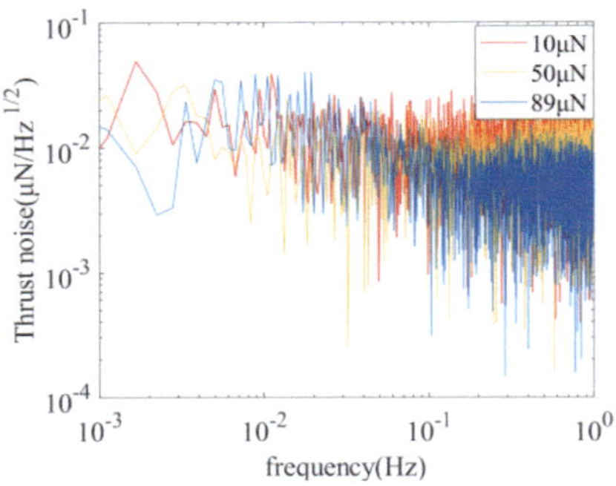

(a) MDIT

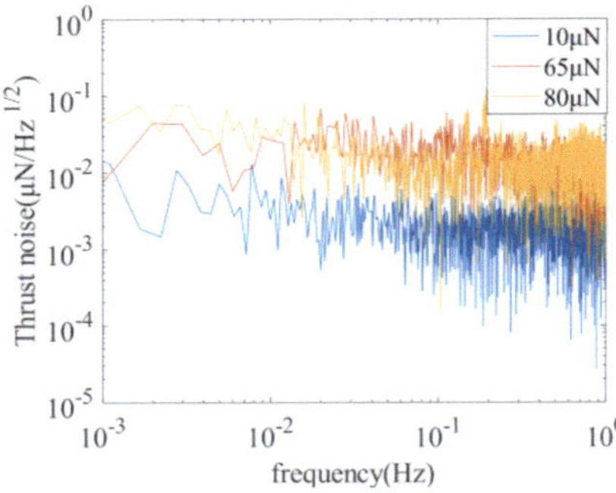

(b) MDCFT

Fig. 18.4 Thrust noise under feedback control system

adjustment ability of the microwave power to the characteristic current is stronger. Besides, we also consider the dynamics of the controlled object under different operations to ensure the stability of the PI controller. The thrust noises of MDIT and MDCFT under the closed-loop conditions are shown as Fig. 18.4. It can be seen that the thrust noises of MDIT and MDCFT are both lower than $0.1\mu N/Hz^{1/2}$ under different thrust levels, satisfying the requirements of space gravitational wave detection. Test results show that the feedback controller can reduce thrust noise to 1% of open-loop thrust noise.

References

1. S.L. Huang, X.F. Gong, P. Xu et al., Gravitational wave detection in space-a new window in astronomy. Phys. Mech. Astronom. **47**, 010404 (2017)
2. M. Armano, H. Audley, J. Baird et al., Beyond the required LISA free-fall performance: new LISA pathfinder results down to 20 μ Hz. Phys. Rev. Lett. **120** (6), 061101 (2018)
3. M. Ando, S. Kawamura, N. Seto et al, DECIGO and DECIGO pathfinder. Cl. Quant. Grav. **27** (8), 084010 (2010)
4. J. Luo, L. Chen, H. Duan et al., TianQin: a space-borne gravitational wave detector. Cl. Quant. Grav. **33**, 035010 (2016)
5. W. Run, Z. Guo, R. Cai, Y. Zhang, Taiji program: gravitational-wave sources. Int. J.Mod. Phys. **35** (17), 2050075 (2020)
6. Armano M, Audley H, Baird J et al., LISA pathfinder platform stability and drag-free performance. Phys. Rev. D **99** (8), 82001 (2019)
7. H. Liu, X. Niu et al., Review of micro propulsion technology for space gravitational waves detection. Acta Astronaut. **193**, 496–510 (2022)
8. G. Anderson et al., Experimental results from the ST7 mission on LISA pathfinder. Phys. Rev. D (98), 102005 (2018)
9. M. Armano, H. Audley, J. Baird et al., Lisa pathfinder micronewton cold gas thrusters: in-flight characterization. Phys. Rev. D **99**, 122003 (2019)
10. S.-Y. Xu, L.-X. Xu, L.-X. Cong et al., First result of orbit verification of Taiji-1 hall micro thruster. Int. J. Mod. Phys. **36** (11&12) (2021)

11. J.-W. He, L. Duan, K. Qi, Ground performance tests and evaluation of RF ion micro-thrusters for Taiji-1 satellite. Int. J. Mod. Phys. **36**, 2140014 (2021)
12. J. Luo, Y. Bai, L. Cai et al., The first round result from the TianQin-1 satellite. Cl. Quant. **37**, 185013 (2020)
13. S. Meng et al., A minimized electron cyclotron resonance ion thruster for China's space-borne gravitational wave detection missions. Cl. Quant. **40**, 175006 (2023)
14. H. Liu, M. Zeng, Z. Chen, L. Qiao, Electron cyclotron resonance discharge enhancement in a cusped field thruster. Plasma Sources Sci. Technol. **30**, 9 (2021)
15. M. Zeng, Research on continuously adjustable micro-newton cusped field Hall propulsion. Harbin Institute of Technology (2023)
16. X. Niu, H. Liu, B. Zhang, Y. Daren, Demonstration of feedback control regula-tion process for microwave discharge ion thruster in space gravitational wave detection. Adv. Space Res. **74**, 5950–5961 (2024)

Chapter 19
Ground Experimental Study of a Drag-Free Satellite based on Air-bearing Testbed

Yankai Wang and Ti Chen

Abstract The drag-free satellite is a critical platform for space gravitational wave detection missions, and the high control accuracy of the drag-free and attitude control system enables the detection missions. Due to the strong dynamic coupling between test masses and satellite, the ground experiment is a fundamental step for the verification of the applied controller. This study presents a Drag-Free Satellite Hardware-in-the-Loop Simulation (DFS-HILS) system based on the planar air-bearing testbed. Moreover, an extended mechanical structure is designed to achieve actuation of the test masses. Finally, the preliminary experimental results show that designed simulation system and the proposed control strategy can successfully realize the drag-free control and the electrostatic suspension.

19.1 Introduction

Gravitational waves (GWs) detection is a verification of A. Einstein's theory and marks the beginning of a new era in physics and astronomy. To detect low-frequency signals of GWs and overcome the limitations of ground-based interferometers, space-borne GW detection is indispensable. The space-borne GW detection missions primarily include TianQin [1], LISA [2], Taiji [3] and so on.

The large-scale space-borne laser interferometer consisting of three drag-free satellites is the primary method for space-borne GW detection. As a crucial platform for detection missions, a drag-free satellite shall be mainly composed of [4]: (1) a satellite body; (2) two movable optical sub-assemblies and (3) two test masses (TMs). Different with traditional satellites, drag-free satellites are characterized by multiple degree-of-freedom (DoF), nonlinearity, and strong coupling. To reduce risks and enhance reliability of the mission, the control method must be initially verified under the ground-based conditions before the costly space detection missions.

Y. Wang · T. Chen (✉)
Nanjing University of Aeronautics and Astronautics, Nanjing, China
e-mail: chenti@nuaa.edu.cn

J. Luo et al. (eds.), *Proceedings of the 7th International Workshop on the TianQin Science Mission; 25–26 April 2024, Hong Kong SAR, China*, Springer Proceedings in Physics 428, https://doi.org/10.1007/978-981-96-7803-7_19

There have been some ground experiments about the space-borne GW detection. The primary method to simulate the micro-gravity environment of the TM is the use of torsion pendulums [5–8]. However, all these studies focused on ground experiments involving the TMs and inertial sensors to valid the performance and index requirements of these modules. Such ground experiments will overlook the strong coupling effects between the various subsystems, making it difficult to valid the high-dimensional dynamics and the responding controllers. Planar air-bearing micro-gravity technology has been widely used for conducting the ground experiments of satellite formation [9–11]. Nevertheless, existing air-bearing testbeds primarily focus on verifying traditional spacecraft dynamics and control, which are not applicable to validating the multi-DoF nonlinear coupling of drag-free satellite.

In order to verify drag-free controller and dynamics coupling, a Drag-Free Satellite Hardware-in-the-Loop Simulation (DFS-HILS) system is built based on air-bearing testbed in this paper. Inspired by [12], the main part of the simulator is designed to simulate the three-DoFs planar motion of a satellite. By mounting the air-bearing in reverse on the top of the simulator body, a microgravity environment for the TM is established.

19.2 Experimental System Architecture

As shown in Fig. 19.1a, the DFS-HILS system consists of the satellite simulators with electrostatic suspension simulation mechanism, a 5×10 m^2 granite testbed, a motion tracking system and a wireless transmission system.

The granite platform provides a frictionless surface for the satellite simulators, while the motion tracking system senses a batch of fluorescent markers attached to the simulator at a certain frequency. The main hardware components of motion tracking system include the motion capture cameras, an image processing workstation, and markers. The workstation acquires the high-precision pose information by tracking

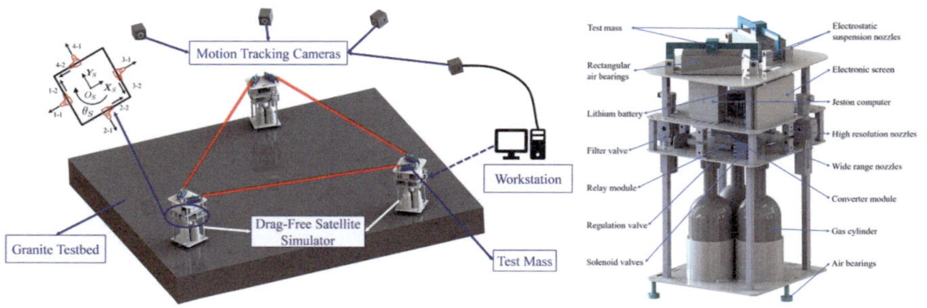

(a) Overview of the Experimental System (b) Simulator Hardware Layout

Fig. 19.1 Schematic of the ground experimental system

markers using array infrared cameras. Furthermore, in order to realize the closed-loop control, the wireless transmission system sends the information to the on-board computer.

As shown in Fig. 19.2, the TMs are placed on the rectangular air bearings, allowing TMs to approximately achieve frictionless motion on a flat surface. Similar to the simulator, the i-th TM has three DoFs: the translational motion along the $\boldsymbol{X}_{TMi}$ and $\boldsymbol{Y}_{TMi}$ directions and the rotational angle θ_{TMi}. The sensitive axis is aligned with the $\boldsymbol{X}_{TMi}$ direction, and the relative displacement of the TM along this axis is controlled by the drive simulator. The non-sensitive axes are the $\boldsymbol{Y}_{TMi}$ direction and θ_{TMi}, with the relative position and attitude controlled by the electrostatic suspension nozzles. Specifically, the control forces and torques along the non-sensitive axes are provided by F_{E1}, F_{E2}, F_{E3} and F_{E4}.

The hardware architecture of a simulator as shown in Fig. 19.1b. The fundamental working principle of the simulator can be found in [12]. Based on the classification of functions, each simulator has three subsystems, namely, the mechanical subsystem, the electronic subsystem and the pneumatic subsystem. As the support structure of the simulator, the components of mechanical subsystem are mainly made by aluminum. The electronic subsystem shown in Fig. 19.3a is mainly responsible for the equipment power supply, data transmission and command issuance during the experiment.

As shown in Fig. 19.3b, the pneumatic subsystem enables the simulator to achieve frictionless motion and jet propulsion on the granite testbed while allowing the TMs to move frictionlessly on the air-bearing, as well as controlling the planar motion of test mass to simulate the electrostatic suspension control. There are three different sets of nozzles, which are electrostatic suspension (ES) nozzles, wide range (WR)

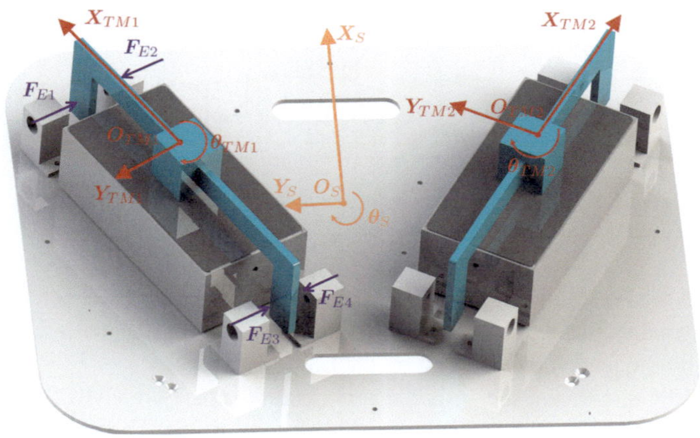

Fig. 19.2 Vertical view of the simulator

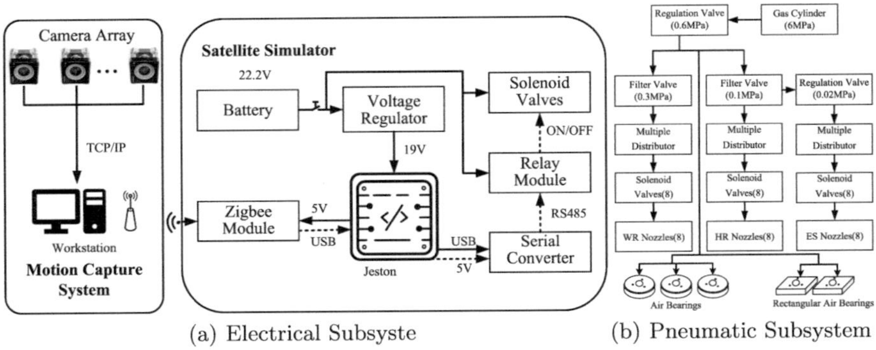

Fig. 19.3 Schematic view of the electrical subsystem and the pneumatic subsystem

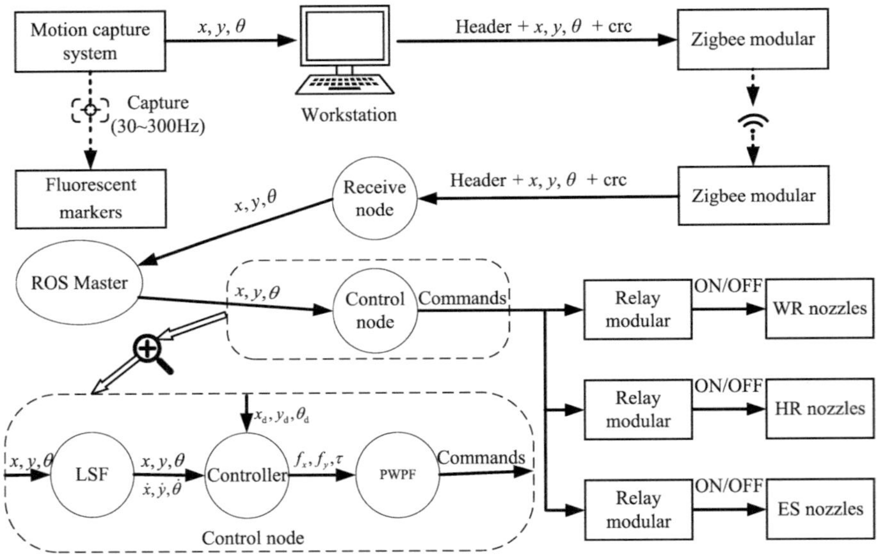

Fig. 19.4 Schematic view of the software architecture

nozzles and high resolution (HR) nozzles. Moreover, an extended mechanical structure is designed to achieve actuation of the TMs based on the electrostatic suspension simulated by the jet force generated by the ES nozzles.

Figure 19.4 shows the the software architecture of the DFS-HILS system. The main frame is built using the robot operating system (ROS). The pulse-width pulse-frequency (PWPF) method is utilized to transform a continuous signal into the switching signal. In the preliminary test experiments, both the drag-free and the electrostatic suspension controllers in Fig. 19.4 are based on proportional derivative control strategies. Key components such as the feedback loop, which incorporates data from motion capture system to adjust actuator commands in real time. The primary objective of the drag-free control is to ensure that the relative displacement of TM along the sensitive axis is ideally kept to zero. And the electrostatic suspension control aims to maintain the relative motion of the TM along the non-sensitive axis at zero. The control laws are given by:

$$\boldsymbol{u}_{DF} = -\boldsymbol{K}_{p1}\boldsymbol{e}_1(t) - \boldsymbol{K}_{d1}\dot{\boldsymbol{e}}_1(t) \tag{19.1}$$

$$\boldsymbol{u}_{ES} = -\boldsymbol{K}_{p2}\boldsymbol{e}_2(t) - \boldsymbol{K}_{d2}\dot{\boldsymbol{e}}_2(t) \tag{19.2}$$

where $\boldsymbol{K}_{p1}$, $\boldsymbol{K}_{d1}$, $\boldsymbol{K}_{p2}$ and $\boldsymbol{K}_{d2}$, are the diagonal positive definite matrices, $\boldsymbol{e}_1(t)$ is the error between the simulator's position and the drag-free reference point, $\boldsymbol{e}_2(t)$ is the relative motion of the TM along the non-sensitive axis.

19.3 Experimental Result

This section presents the preliminary ground experimental results with single TM. Note that the measurement accuracy of the motion capture system is about 0.2 mm. The time history of the satellite simulator's position and attitude are indicated in Fig. 19.5a. Both the displacement of the TM relative to the satellite body along the sensitive axis $\boldsymbol{X}_{TM1}$ and the relative motion along insensitive axes $\boldsymbol{Y}_{TM1}$ and θ_{TM1} are given in Fig. 19.5b. The red dashed lines represent the upper and lower bounds of

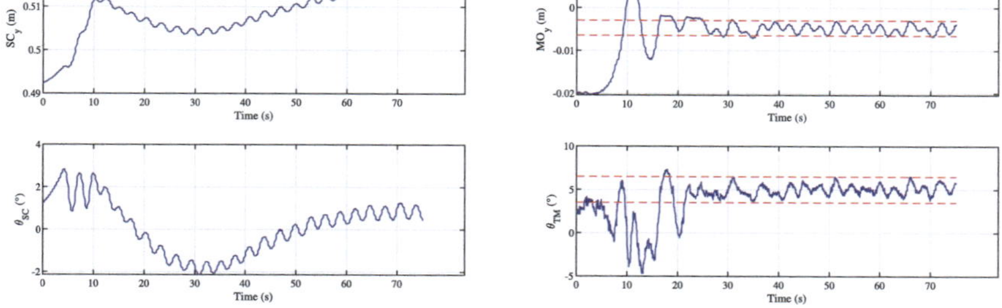

(a) The position of the simulator (b) The relative motion error of TM

Fig. 19.5 Drag-free control and electrostatic suspension control results

the fluctuations in relative motion in the stabilization stage. As shown in Fig. 19.5a, $\boldsymbol{X}_{TM1}$ converges to zero at about 20 s, and the convergence error is less than ± 5 mm. After 30s, the relative movement along $\boldsymbol{Y}_{TM1}$ and the relative attitude error θ_{TM1} are less than 5mm and $5°$, respectively.

19.4 Conclusions

In this paper, a ground experimental system based on the planar air-bearing testbed is designed for drag-free satellite. The system can simultaneously simulate the planar motion of the satellite and two TMs, offering 9 DoFs. The mechanical, electronic and pneumatic subsystems are developed and tested. To simulate the electrostatic suspension, a specially designed additional structure for the TM is presented. A preliminary experiment with a single TM is conducted to verify the drag-free control and the electrostatic suspension control. Experimental results demonstrate that the designed DFS-HILS system can successfully simulate the science mode of the drag-free satellite.

Acknowledgements This work was supported by the National Key Research and Development Program of China (Grant No. 2022YFC2204800).

References

1. J. Luo et al., TianQin: a space-borne gravitational wave detector. Classical Quant. Gravity **33**(3), 035010 (2016)
2. K. Danzmann, LISA Study Team, et al. Lisa: laser interferometer space antenna for gravitational wave measurements. Classical Quantu. Gravity **13**(11A), A247 (1996)
3. Z. Luo, Y. Wang, Y. Wu, W. Hu, G. Jin, The Taiji program: A concise overview. Progr. Theor. Exp. Phys. **2021**(5), 05A108 07 (2020)
4. Y. Fang, X. Zhang, F. Fangyuan, H. Li, Payload architecture and pointing control strategies for tianqin. Phys. Rev. D **109**(6), 062001 (2024)
5. L. Carbone, A. Cavalleri, R. Dolesi, C.D. Hoyle, M. Hueller, S. Vitale, W. Joseph Weber, Achieving geodetic motion for lisa test masses: ground testing results. Phys. Rev. Lett. **91**(15), 151101 (2003)
6. G. Russano, *A Torsion Pendulum Ground Test of the LISA Pathfinder Free-fall Mode* PhD thesis (University of Trento, 2015)
7. H.B. Tu, Y.Z. Bai, Z.B. Zhou, L. Liu, L. Cai, J. Luo, Performance measurements of an inertial sensor with a two-stage controlled torsion pendulum. Classical Quant. Gravity **27**(20), 205016 (2010)
8. J. Lin, J. Lian, G. Zhao, H. Li, X. Chi, A 2-dof drag-free control ground simulation system based on stewart platform. ISA Trans. **146**, 528–540 (2024)
9. T. Rybus, K. Seweryn, Planar air-bearing microgravity simulators: Review of applications, existing solutions and design parameters. Acta Astronaut. **120**, 239–259 (2016)

10. Z. Huang, L. Yun, H. Wen, D. Jin, Ground-based experiment of capturing space debris based on artificial potential field. Acta Astronaut. **152**, 235–241 (2018)
11. Z. Wei, T. Chen, H. Wen, D. Jin, H. Haiyan, Experimental study on autonomous assembly of multiple spacecraft simulators in a spinning scenario. Acta Astronaut. **207**, 106–117 (2023)
12. Z. Huang, W. Zhang, T. Chen, H. Wen, D. Jin, Characterizing an air-bearing testbed for simulating spacecraft dynamics and control. Aerospace **9**(5) (2022)

Chapter 20
Low-Frequency High-Precision and Large Bandwidth Quadrant Photodetector for Space-Based Gravitational Wave Detection

Liu-Yang Chen, Zhi-Lin Xu, Yuan-Ze Jiang, Yu-Jie Feng, and Yu-Rong Liang

Abstract In this work, a low-frequency high-precision quadrant photodetector (QPD) with large bandwidth was studied for transponder type inter-satellite laser heterodyne interferometer of space-based gravitational wave detection. By conducting the in-depth theoretical modeling and noise analysis on the QPD, suppression strategies for the key noise sources including photodiode junction capacitance and operational amplifiers were proposed. For experimental demonstration, the preliminary QPD was fabricated and tested. The response frequency from 5 to 25 MHz, and the phase noise floor of 10 μrad/Hz$^{1/2}$ at 6 mHz were experimentally achieved by the fabricated QPD, which were close to the stringent phase reading requirement of the TianQin mission.

20.1 Introduction

Compared with ground-based gravitational wave (GW) detection, space-based GW detection takes advantage of the immunity to terrestrial gravity gradient noise and the significantly larger effective arm length of the detector which facilitates to extend the detection frequency range down to the millihertz (mHz) regime [1]. Transponder type inter-satellite laser heterodyne interferometer is an integral part of GW detectors because of its ability to conquer the problems of Doppler shifts, long-distance laser propagation and the inevitable beam divergence which lead to weak interferometric signals [2]. According to the detection goals proposed by TianQin, the displacement

L.-Y. Chen · Z.-L. Xu (✉) · Y.-Z. Jiang · Y.-J. Feng · Y.-R. Liang
National Gravimetry Laboratory, MOE Key Laboratory of Fundamental Physical Quantities Measurement, and School of Physics, Huazhong University of Science and Technology, Wuhan, China
e-mail: zlxu_phys@hust.edu.cn

J. Luo et al. (eds.), *Proceedings of the 7th International Workshop on the TianQin Science Mission; 25–26 April 2024, Hong Kong SAR, China*, Springer Proceedings in Physics 428, https://doi.org/10.1007/978-981-96-7803-7_20

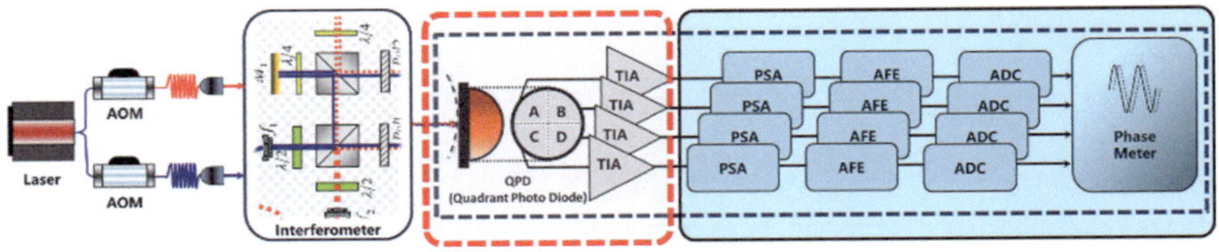

Fig. 20.1 Schematic of the optical phase read system

sensitivity of transponder type inter-satellite laser heterodyne interferometer should reach the picometer (pm) level [3].

Transponder type inter-satellite laser heterodyne interferometer is composed of heterodyne interferometer, quadrant photodetector (QPD), and phasemeter. Particularly, QPD enclosed in red dashed lines in Fig. 20.1 serves as a bridge connecting the heterodyne interferometer with the phasemeter, thus holding important research value. QPD research still faces challenges because the displacement sensitivity of transponder type inter-satellite laser heterodyne interferometer reaches the pm level.

20.2 Theoretical Analysis

QPD includes a photodiode (PD) for converting optical signals to currents and a trans-impedance amplifier (TIA) circuit for converting currents to voltages signals. The schematic diagram of the QPD based on the operational amplifier (Op-Amp) is shown in Fig. 20.2.

The electrical circuit of the QPD can be considered as a special-purpose version of TIA circuit. It can be equivalent to a second-order closed-loop transfer function and the unipolar point model can be used to fit the open-loop gain curve of non-ideal Op-Amp. After circuit analysis, the transfer function of the QPD can be obtained [4]:

Fig. 20.2 Schematic diagram of the QPD based on Op-Amp

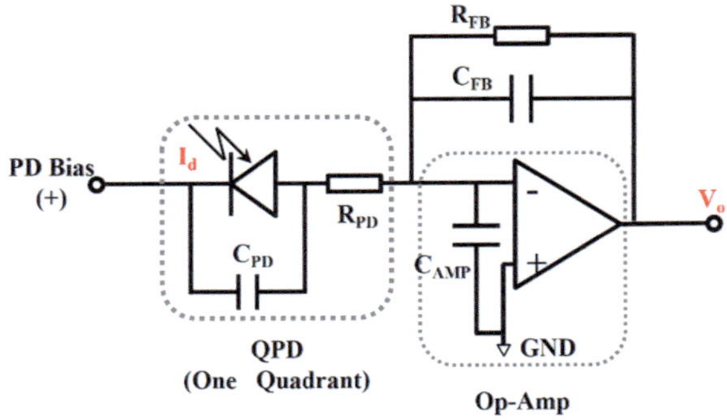

$$TF(\omega) = \frac{V_o}{I_d} = R_{FB}\frac{A_{ol}}{A_{ol}+1}\frac{\omega_0^2}{(j\omega)^2 + \frac{j\omega\omega_0}{Q} + \omega_0^2}. \tag{20.1}$$

where V_o and I_d are respectively the input and output of the QPD. A_{ol} is open-loop gain of Op-Amp at the flat region of the transfer function. ω is the operating frequency of the circuit. $\omega_0 = \sqrt{\frac{\omega_A(A_{ol}+1)}{R_{FB}(C_{pd}+C_{FB})}}$ is the intrinsic frequency of the circuit, and ω_A is frequency of Op-Amp at which the open-loop gain is $-$ 3 dB. $Q = \frac{\omega_0}{\omega_A\left(1+A_{ol}\frac{C_{FB}}{C_{pd}+C_{FB}}\right)+\frac{1}{R_{FB}(C_{pd}+C_{FB})}}$ is the quality factor of the circuit [5]. C_{pd} is the junction capacitance of PD. R_{FB} and C_{FB} are respectively the feedback resistance and feedback capacitance of Op-Amp.

The transfer function describes the relationship between the input and output of a linear time-invariant system. The magnitude frequency response is equal to the modulus of the transfer function: Gain(ω) $= abs(TF(\omega))$. As the frequency increases, the voltage the magnitude frequency response decreases. The QPD bandwidth is defined as the frequency value at which the magnitude frequency response drops to 0.707 times of its maximum value, corresponding to a power attenuation of 3 dB. Due to the Doppler frequency drift effect between the two spacecraft in space-based GW detection, the QPD bandwidth must exceed 25 MHz [2].

From Eq. (20.2) and the expression of ω_0 and Q, it can be seen that C_{pd} and C_{FB} are the two factors influencing the QPD bandwidth. Since C_{pd} is usually more than ten times of C_{FB}, the effect of the junction capacitance C_{pd} on the QPD bandwidth is significantly greater than that of the feedback capacitance C_{FB}. Therefore, only C_{pd} is considered in the following discussion.

The junction capacitance C_{pd} should be considered in the light of the noise contribution and the requirements from differential wavefront sensing (DWS), used for transponder type inter-satellite laser heterodyne interferometer pointing control. According to the noise model, the smaller C_{pd} is in favor of achieving the lower noise level. However, from the perspective of DWS, the photosensitive surface size of the QPD should be sufficiently large to increase the measurement sensitivity, which, however, will increase the junction capacitance of the QPD. Therefore, the junction capacitance C_{pd} of the QPD can't be satisfactory small, usually around 10 pF.

Op-Amp's parameters are also the factors influencing the QPD bandwidth. Typically, gain bandwidth product (GBP) defined by GBP $= (A_{ol}+1)\omega_A / 2\pi$ is used to characterize the comprehensive effects of A_{ol} and ω_A of Op-Amp.

The effects of the junction capacitance C_{pd} and GBP on the QPD bandwidth are calculated and presented in Fig. 20.3. It is observed that reducing the junction capacitance C_{pd} and increasing GBP can boost the QPD bandwidth. When the junction capacitance is 10 pF and GBP is greater than 0.4 GHz, the QPD bandwidth can reach 25 MHz. Hence, selecting an Op-Amp with a high GBP that also satisfies the requirements for noise analysis as outlined below is crucial in the fabrication of the QPD.

The phase noise of the QPD can be analyzed using the principle of small-signal vector superposition since the noise is generally much smaller than the strength of

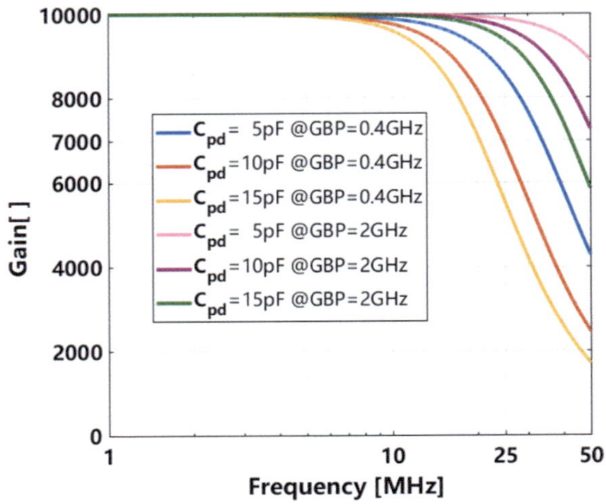

Fig. 20.3 Effect of junction capacitance on the QPD bandwidth

the interference signal. As a reference, the carrier-to-noise ratio of the QPD is usually better than 40 dB. The phase noise of the detector can be represented as [6]:

$$\varphi_{\text{noise}} = \sqrt{V_{OA}\left(1/R_{FB}^2 + \omega^2 C_{pd}^2\right) + I_{OA}/I_{AC}}. \tag{20.2}$$

where V_{OA} is the equivalent input voltage noise of Amp-Op. I_{OA} is the equivalent input current noise of Amp-Op. I_{AC} is the photocurrent generated by the heterodyne interference signal which can be treated as a constant in noise analysis. Equation (3) reveals that the phase noise is mainly influenced by C_{pd}, V_{OA}, and I_{OA}. The effects of the junction capacitance C_{pd} as well as V_{OA}, and I_{OA} of Op-Amp on the phase noise of the QPD are illustrated in Fig. 20.4. It is observed that the effects of C_{pd} and V_{OA} are coupled and balanced with each other, requiring a comprehensive consideration of their performance. The requirement of equivalent current noise I_{OA} is relatively clear. Considering a phase noise contribution limit of 1 μrad/Hz$^{1/2}$, I_{OA} is required to be less than 2.6 pA/Hz$^{1/2}$ at 5 ~ 25 MHz.

20.3 Experimental Setup and Experimental Result

After noise analysis in Sect. 20.2, the key parameters of the QPD are designed as $C_{PD} = 10$ pF(bais = 5V), $V_{OA} = 1\text{nV/Hz}^{1/2}$ at 5 ~ 25 MHz, $I_{OA} = 2.5\text{pA/Hz}^{1/2}$ at 5 ~ 25 MHz, GBP = 1.8GHz.

For experimental demonstration, a preliminary version of the QPD was fabricated based on the designed parameters, as shown in the right part of Fig. 20.5. An $LiNbO_3$ intensity modulator is used to generate the interference signal to test the phase noise of the fabricated QPD, as shown in the left part of Fig. 20.5. A Mach–Zehnder

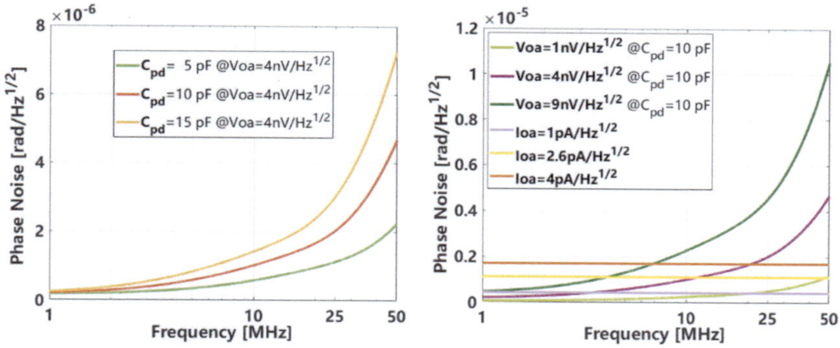

(a). Effect of junction capacitance on the phase noise of the QPD

(b). Effect of Op-Amp on the phase noise of the QPD

Fig. 20.4 Analysis of the contribution of electrical parameters on the phase noise of the QPD

interferometer (MZI) is contained in the $LiNbO_3$ intensity modulator. Since the arm length difference is less than 3.1 μm, the MZI in the $LiNbO_3$ is insensitive to the frequency noise of the laser, making it naturally suitable for using use as a QPD test platform. Phase demodulation employs a digital phase-locked loop based on the FPGA platform, which achieves a differential phase measurement between channels better than 3 μrad/Hz$^{1/2}$ above 6 mHz. To isolate the impact of laser and phasemeter on the phase noise testing of QPD, Pilot-Tone correction was applied to minimize clock jitter noise. Optical power stabilization was applied to reduce amplitude to phase noise coupling. The experimental setup was placed in a polyethylene foam box to isolate the air disturbances and temperature fluctuations from the laboratory.

The noise floor of the fabricated QPD is characterized by the phase amplitude spectral density (ASD), as shown in Fig. 20.6. It can be seen that the noise floor of optical phase above 6 mHz is down to 10 μrad/Hz$^{1/2}$. A positive correlation between phase noise and the frequency of the heterodyne interference signal can be seen in Fig. 20.6 which is consistent with the theoretical derivation trend in Fig. 20.4.

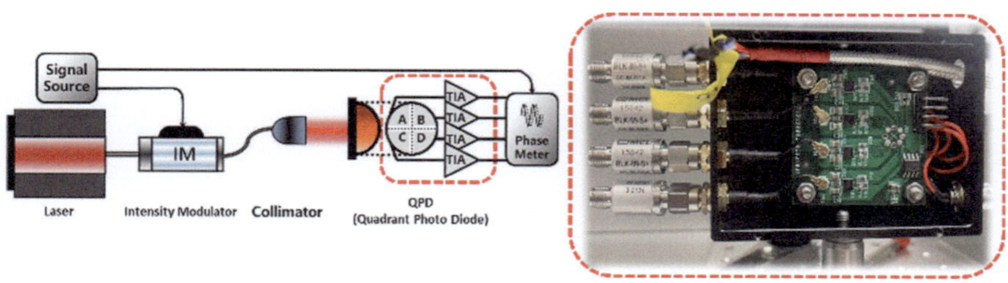

Fig. 20.5 The optical path of the intensity modulator and a photograph of the QPD used to test the noise

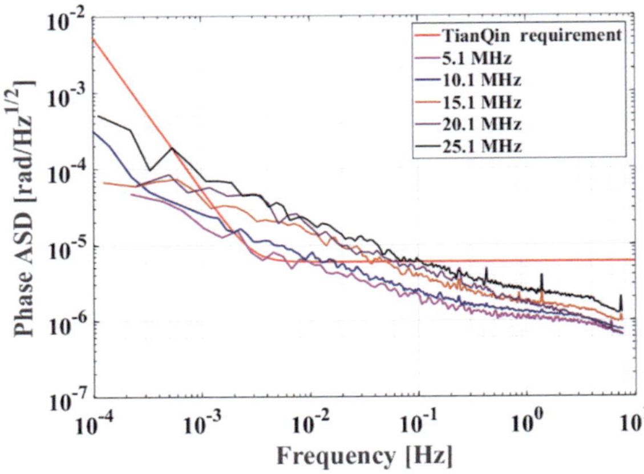

Fig. 20.6 Phase ASD of the QPD noise floor at different heterodyne frequencies

However, the noise floor and frequency values do not exhibit a perfect correspondence. Further studies are necessary to reveal whether the phase noise floor is limited by the light impurities of the LiNbO3 or the electronic noise of the QPD.

20.4 Summary

The QPD for transponder type inter-satellite laser heterodyne interferometer of space-based GW detection was studied theoretically and experimentally. By conducting the in-depth theoretical modeling and noise analysis on the QPD, suppression strategies for the key noise sources including the photodiode junction capacitance of the QPD and the Op-Amp were proposed. An initial of QPD was fabricated and then tested via a $LiNbO_3$ intensity modulator setup, the phase noise floor of 10 μrad/Hz$^{1/2}$ at 6 mHz at heterodyne frequency from 5 to 25 MHz was achieved experimentally.

The next research will focus on refining modulation approaches and noise mitigation strategies in the QPD. An ultra-stable optical platform is planned to construct to substitute the current intensity modulator scheme, validating the contribution of the QPD on the noise floor of optical phase reading system in a more authentic way.

References

1. Y. Gong, J. Luo, B. Wang, Concepts and status of Chinese space gravitational wave detection projects. Nat. Astron. **5**(9), 881–889 (2021)
2. H.Z. Duan, Y.X. Luo, J.Y. Zhang et al., Inter-satellite laser interferometry. Acta Sci. Natur. Univ. Sunyatseni **60**(1–2), 162–177 (2021)
3. J. Luo, L.S. Chen, H.Z. Duan et al., TianQin: a space-borne gravitational wave detector. Cl. Quantum Grav. **33**(3), 035010 (2016)

4. X. Ramus, Transimpedance considerations for high-speed amplifiers. Application Report SBOA122. Texas Instruments 97 (2009)
5. Y.Z. Jiang, X.L. Jin, H.C. Yeh et al., Shot-noise-limit performance of a weak-light phase readout system for intersatellite heterodyne interferometry. Opt. Express **29**(12), 18336–18350 (2021)
6. A. Joshi, J. Rue, S. Datta, Low-noise large-area quad photoreceivers based on low-capacitance quad InGaAs photodiodes. IEEE Photonics Technol. Lett. **21**(21), 1585–1587 (2009)

Chapter 21
Progress on Phase Measurement Technique for TianQin

Yuan-Ze Jiang, Yu-Jie Feng, Liu-Yang Chen, Yu-Jie Tan, Ming-Yang Xu, and Yu-Rong Liang

Abstract The TianQin project plans to send three satellites into Earth orbit, forming a triangular array with sides about 170,000 km long. Laser interferometers will be set up to measure the relative motions of the test masses. The overall displacement resolution of the optical metrology system needs to achieve 1 pm/Hz$^{1/2}$@6 mHz. The performance of the phasemeters extracting the interferometric phases is one of the baselines to fulfill the requirement of the accuracy of the displacement measurement. In this paper the core metrology of the phasemeter is introduced and the performance of the 4-channel phasemeter is tested using signal generator. After improving temperature stabilities around the phasemeter and the signal generator, the differential phase noise of the phasemeter prototype achieved 1 μrad/Hz$^{1/2}$@6 mHz.

21.1 Background

TianQin [1, 2] is a space-borne detector orbiting the Earth to measure gravitational waves in the millihertz frequencies with designed accuracy of 1 pm/Hz$^{1/2}$@6 mHz. Three satellites form a triangular array with sides about 170,000 km long to measure the relative movements between 6 test masses inside the satellites by laser interferometers (Fig. 21.1).

The interferometers transfer tiny movements of TMs to phase variations of interference light field monitored by phasemeters based on digital phase-locked loop (DPLL). Three sets of laser interferometric data are needed to measure the relative displacement of each pair of TMs, and each phase signal is obtained via laser interference, photoelectric conversion, pilot signal superposition, phase-locked loop and pilot correction (Fig. 21.2).

Y.-Z. Jiang · Y.-J. Feng · L.-Y. Chen · Y.-J. Tan · M.-Y. Xu · Y.-R. Liang (✉)
MOE Key Laboratory of Fundamental Physical Quantities Measurement and Hubei Key Laboratory of Gravitation and Quantum Physics, PGMF and School of Physics, Huazhong University of Science and Technology, Wuhan, China
e-mail: liangyurong20@hust.edu.cn

J. Luo et al. (eds.), *Proceedings of the 7th International Workshop on the TianQin Science Mission; 25–26 April 2024, Hong Kong SAR, China*, Springer Proceedings in Physics 428, https://doi.org/10.1007/978-981-96-7803-7_21

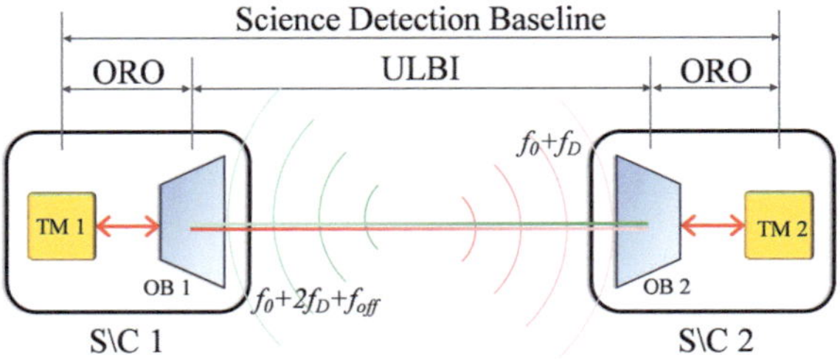

Fig. 21.1 Piecewise measurement configuration of a pair of TMs locating in separate satellites. ORO: optical readout system for test mass. ULBI: ultra-long base line interferometry. TM: test mass. OB: optical bench (including telescope). S\C: space craft

Fig. 21.2 Schematic of interferometric phase measurement

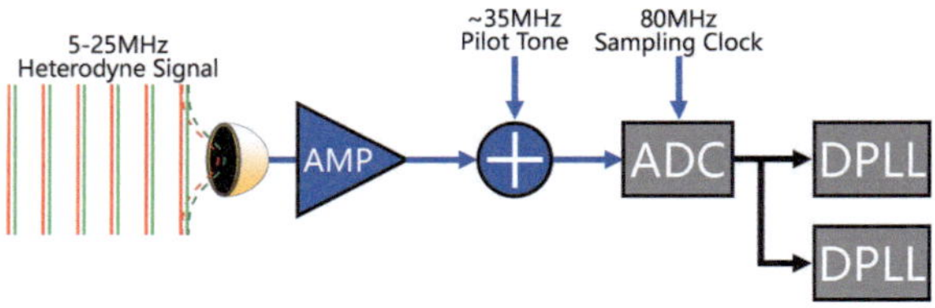

All the measurement data need to be combined according to the time-delay interferometry to obtain the final measurement result. In order to meet the final measurement level, the precision of phase measurement needs to be further improved. In this paper, the measurement principle of phasemeter based on DPLL is introduced, and the current experimental results in our group are exhibited.

21.2 Measurement Principle and Theoretical Analysis

In intersatellite laser interferometry, the main tasks of the phasemeter include converting the analog output of the photodetector to digital signal, real-time tracking of signal phase, pointing control of the laser beam and heterodyne phase locking. The Field Programmable Gate Array (FPGA) chip is used as the hardware core to build a phasemeter, which can realize synchronous acquisition and phase readout of multiple RF signals (Fig. 21.3).

Inside the DPLL, the input signal is first mixed with a sinusoidal signal of the same frequency (or close-in frequency) generated by numerically controlled oscillator (NCO) using a look-up table (LUT), and the low frequency signal obtained by mixing is fed back to the NCO. NCO uses a phase increment register (PIR) and a phase accumulator (PA) to correct its output frequency in real time. Through this process, real-time tracking of input signal frequency and real-time output of frequency and phase can be realized. In order to achieve the measurement accuracy,

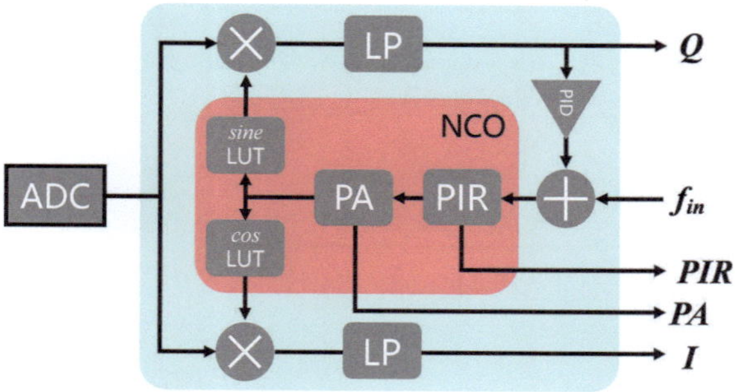

Fig. 21.3 Diagram of the phasemeter based on ADPLL

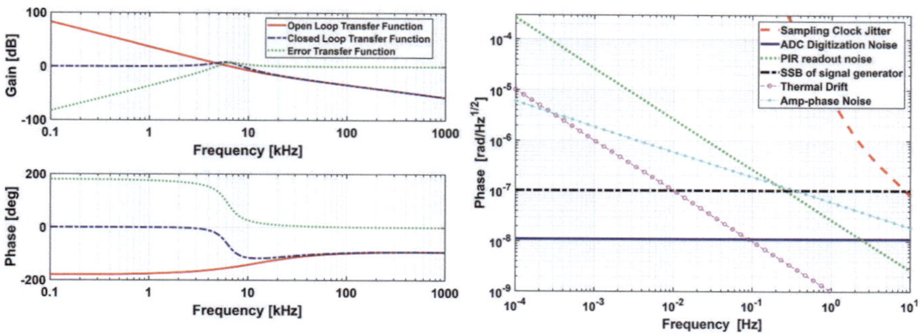

Fig. 21.4 Transfer function of the DPLL (left) and expected noises of the phasemeter (right). In our simulation, PIR readout noise, Amp-phase noise and thermal drift are expected to be main noise sources of the phase measurement process

the pilot correction strategy must be applied, and the stabilities of analog circuit, clock stability and other aspects must be seriously considered.

The transfer function of the DPLL [3–5] and the expected noises of the phasemeter are presented in Fig. 21.4. In our simulation, PIR readout noise, Amp-phase noise and thermal drift are expected to be main noise sources of the phase measurement process [6].

21.3 Experimental Setup and Result

Our experimental structure diagram is shown in the Fig. 21.5. We use a dual-channel signal generator to produce signals of 10.1 MHz and 34.1 MHz, presenting the heterodyne signal from QPR and the pilot tone signal. The superposed signal is split to four identical parts by power splitter and transferred to four SMA inputs of the phasemeter installed on a commercial FPGA platform. The commercial FPGA platform provides a daughter card with two identical ADC input channels on one card, hence we can use a pair of daughter cards to build a 4-channel phasemeter.

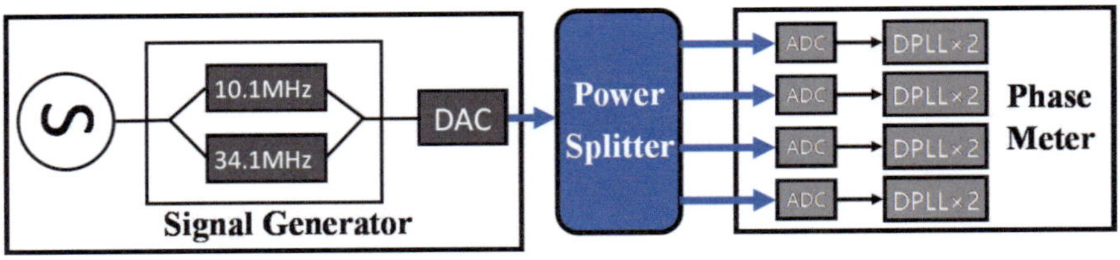

Fig. 21.5 Diagram of the phasemeter test

Fig. 21.6 Image of the test environment

To test the influence of temperature to phase measurement, we built a temperature-controlled box using polyfoam and TEC cooler. After serial tests, we choose to place the signal generator in the temperature-controlled box meanwhile the phasemeter is covered by a paper box for passive thermal isolation to get a better phase noise floor as shown in the Fig. 21.6. The temperature noise floor inside the temperature-controlled box reached 20 $\mathrm{mK/Hz^{1/2}}$@6 mHz while inside the passive thermal isolation paper box reached 200 $\mathrm{mK/Hz^{1/2}}$@6 mHz.

Figure 21.7 shows the phase noise floor measurement results of our test devices. The four solid lines represent single channel phases after pilot tone correction, the dashed and dash dotted lines beneath the solid lines are differential results of the corresponding single channels and the orange thick dash line represents the TianQin length measurement requirement of $1\ \mathrm{pm/Hz^{1/2}} \times [1 + (6\ \mathrm{mHz/f})^4]^{1/2}$.

The results show that after pilot tone correction, the phase noise floors of single channel data and differential data are lower than TianQin length measurement requirement of $1\ \mathrm{pm/Hz^{1/2}} \times [1 + (6\ \mathrm{mHz/f})^4]^{1/2}$. Two channels on the same daughter card show a better common-mode rejection performance, achieving 1 $\mathrm{\mu rad/Hz^{1/2}}$@6 mHz whereas differential results of channels on different daughter cards barely show common-mode suppression effect, with possible reasons including temperature gradient, pilot tone correction residue and reference voltages inconformity of different daughter cards.

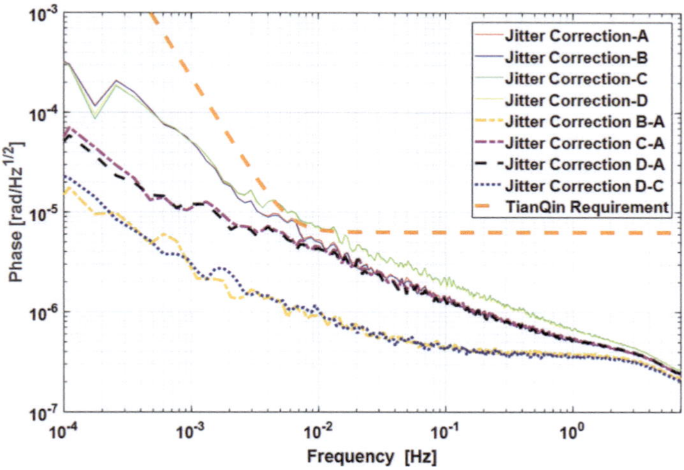

Fig. 21.7 Phase noise floor results. The results show that the best differential phase noises achieved 1 μrad/Hz$^{1/2}$ @6 mHz after pilot tone correction

21.4 Summary

In this paper, we introduce the progress on sub-μrad level phase measurement technique for TianQin project. The measurement principle and theoretical analysis are reviewed and the current experimental results in our group are exhibited. The results show that the differential measurement performance of the phases achieved 1 μrad/Hz$^{1/2}$ @6 mHz on a commercial FPGA platform with proper temperature control of the signal generator and the phasemeter. Meanwhile the experiment exposed several issues needed to be further considered including temperature response of RF source, pilot tone correction residue, thermal uniformity of hardware and so on.

References

1. J. Luo et al., TianQin: a space-borne gravitational wave detector. Class. Quantum Gravity **33**, 35010 (2016)
2. J. Mei et al., The TianQin project: current progress on science and technology. Prog. Theor. Exp. Phys. **2021**, 5A107 (2021)
3. Y.-R. Liang, H.-Z. Duan, H.-C. Yeh, J. Luo, Fundamental limits on the digital phase measurement method based on cross-correlation analysis. Rev. Sci. Instrum. **83**, 95110 (2012)
4. O. Gerberding, B. Sheard, I. Bykov, J. Kullmann, J.J.E. Delgado, K. Danzmann, G. Heinzel, Phasemeter core for intersatellite laser heterodyne interferometry: modelling, simulations and experiments. Class. Quantum Gravity **30**, 235029 (2013)
5. Y.-R. Liang, H.-Z. Duan, X.-L. Xiao, B.-B. Wei, H.-C. Yeh, Note: inter-satellite laser range-rate measurement by using digital phase locked loop. Rev. Sci. Instrum. **86**, 16106 (2015)

6. Y.-R. Liang, Y.-J. Feng, G.-Y. Xiao, Y.-Z. Jiang, L. Li, X.-L. Jin, Experimental scheme and noise analysis of weak-light phase locked loop for large-scale intersatellite laser interferometer. Rev. Sci. Instrum. **92**, 124501 (2021)

Chapter 22
Ground Experimental Demonstration of Intersatellite Laser Interferometry

Fangjie Liao, Huizong Duan, Xiangqing Huang, Jingyi Zhang, Fan Zhu, and Hsien-Chi Yeh

Abstract Inter-satellite laser interferometry is an advanced technique for high-precision measurements in space. In this work we discuss a ground experimental demonstration of transponder type intersatellite laser interferometry. The digital phase-locked loop (DPLL) architecture is used to accurately maintain a 10 MHz frequency difference in the heterodyne interferometer. The displacement measurement accuracy of the interferometer is better than 10 $\mathrm{nm}/\sqrt{\mathrm{Hz}}$ at 0.1 Hz, and the Angle measurement accuracy is better than 0.3 $\mu\mathrm{rad}/\sqrt{\mathrm{Hz}}$ at 0.1 Hz, after the performance optimization. This research provides fundamental data and technical support for further improving the accuracy of optical interferometry.

22.1 Introduction

The Space Advanced Gravity Measurements (SAGM) mission is a satellite gravity project proposed in China, aimed at measuring Earth's gravity field [1]. Another objective is verify the key technology for TianQian mission in future. Inter-satellite laser interferometry , a crucial techniques, involves establishing a laser interferometer between two satellites to measure the relative displacement changes between them. Laser ranging interferometry system is engineered to maintain an intersatellite distance of approximately 200 kilometers and boasts a displacement measurement resolution of 10 $\mathrm{nm}/\sqrt{\mathrm{Hz}}$ at 0.1 Hz. Many technologies are undergoing deeper exploration in scientific missions based on intersatellite laser interferometry, such as GRACE FO(Gravity Recovery and Climate Experiment Follow-On) [2], TianQin [3] and LISA [4].

F. Liao · H. Duan (✉) · X. Huang · J. Zhang · F. Zhu · H.-C. Yeh
MOE Key Laboratory of TianQin Mission, TianQin Research Center for Gravitational Physics and School of Physics and Astronomy, Frontiers Science Center for TianQin, CNSA Research Center for Gravitational Waves, Sun Yat-sen University (Zhuhai Campus), Zhuhai, People's Republic of China
e-mail: duanhz3@mail.sysu.edu.cn

J. Luo et al. (eds.), *Proceedings of the 7th International Workshop on the TianQin Science Mission; 25–26 April 2024, Hong Kong SAR, China*, Springer Proceedings in Physics 428, https://doi.org/10.1007/978-981-96-7803-7_22

To identify potential main noise sources and develop suppression methods for future SAGM missions, a heterodyne interferometer with a 4-meter baseline was constructed to validate the measurement scheme. In this process, we also aim to optimize the spatial layout of the interferometer and develop a more efficient optical path design. In this paper, we provide a detailed introduction to the experimental device and present preliminary measurement results regarding the interferometer's accuracy.

22.2 Theory

The basic idea of inter-satellite laser interferometry is expressed schematically in Fig. 22.1, which represents two spacecraft separated by a large distance. The most promising technology remains the transponder scheme. The laser beam is collimated upon injection, and through the action of the beamsplitter (BS), 5% of it serves as a local reference beam in the first spacecraft (SC_1), while the remaining 95% is transmitted to the distant SC_2. A series of neutral density filters are placed in the transmission path of the beam between the two satellites ,than an extremely faint beam of light is received on the spacecraft 2 (SC_2) , like a light beam traveling a long distance between the satellites.

This laser interferometer will be set up as an optical transponder, and the phase $\overline{\phi}$ of the heterodyne signal is obtained from the phase meter on the SC_1 . The change in optical path length is twice the change in the distance between the satellite simulators. The processing of the phase signal can not only be obtained the relative optical-path-length change between the two interfering beams but also the angle between

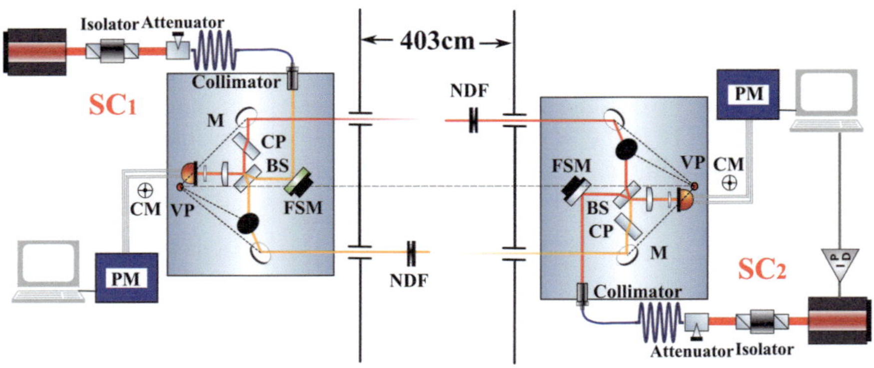

Fig. 22.1 Block diagram of inter-satellite laser interferometry. Yellow lines indicate the laser beam from the first spacecraft(SC_1) and red lines is the laser beam from the seconed spacecraft(SC_2). BS: A beam splitter with a spectral ratio of R : T = 95 : 5, M: Mirror, CP: Counteracting plate, FSM: Fast steering mirror, VP: vertex point of the triple mirror assembly, CM: Spacecraft's center of mass, PM: Phase metter, NDF: Neutral Density Filters

their wavefronts. The longitudinal pathlength signal is derived from the heterodyne signal's phase, where changes in phase are directly interpreted as variations in the propagated pathlength of either beam. We express it by the following correlation relation

$$\delta x = \frac{\lambda}{2\pi}\frac{\overline{\phi}}{2} \tag{22.1}$$

The standard technique to achieve this angular measurement is differential wavefront sensing (DWS), which uses a quadrant photodiode (QPD) to detect the interference pattern. The DWS signal originates from the horizontal or vertical misalignment between the measurement and reference beam axes on the QPD, described by an angle α or β, respectively. The vertical misalignment is illustrated as an example in Fig. 22.2.

$$DWS_{hor} = \frac{(\phi_A + \phi_D) - (\phi_B + \phi_C)}{2} \tag{22.2}$$

$$DWS_{ver} = \frac{(\phi_A + \phi_B) - (\phi_C + \phi_D)}{2} \tag{22.3}$$

For small relative tilt angles, yaw (α) and pitch (β), between the wavefronts of two interfering beams, as evaluated by Differential Wavefront Sensing (DWS), they are expressed as:

$$\alpha \approx \frac{3\lambda}{16r} DWS_{hor}. \tag{22.4}$$

$$\beta \approx \frac{3\lambda}{16r} DWS_{ver}. \tag{22.5}$$

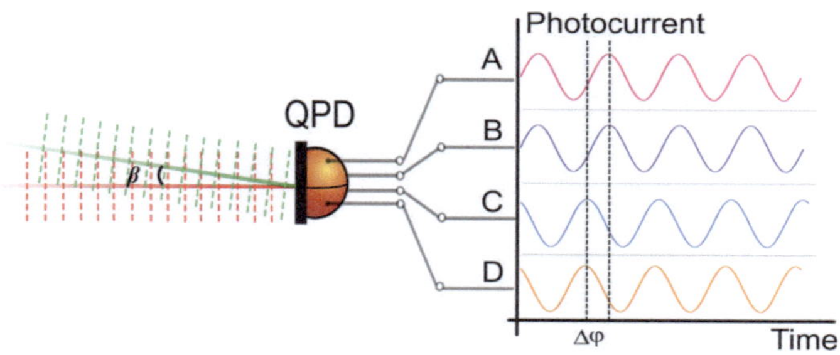

Fig. 22.2 Differential Wavefront Sensing(DWS) principle. Two wave simping on the surface of a quadrant photodiode (QPD) at an angle, and different QPD segments detect different phases

22.3 Experiment

Replicating the vacuum conditions requires vacuum chambers to simulate the low-pressure environment. So experiments are conducted in a vacuum to simulate space conditions, thereby eliminating gas interference with the results, as show in Fig. 22.3. The main interferometer consists primarily of two optical benches and several neutral density filters which are utilized to simulate the weakening of received light. Hydroxide catalysis bonding (HCB) technique is used to assemble the optical components, ensuring strong, stable, and precisely aligned connections on the ultra-stable optical bench [5]. The optical bench (OB) incorporates a corner-cube prism thereby enabling precise alignment and measurement of the beam. The imaging system compresses the beam size, allowing for increased light power utilization in detection. what's more, the axis of emitted beam directed by a fast-steering mirror (FSM) to improve the optical interferometer.

The light from each NPRO is fed to the optical bench of the vacuum chamber by 3 m polarization maintaining fiber feedthroughs. As shown in the Fig. 22.4, the laser beam is incident from the collimator onto the ultrastable optical bench. The fast steering mirror, incorporated into the optical bench (OB), is used to achieve precise control of the beam direction and ensure accurate beam incidence. A neutral density filter is used to adjust the variation in light power during beam transmission.

The heterodyne signal, formed by a weak beam of light and a relatively strong light convert into a proportional output voltage using a high-gain detector. Then, a digital phase-locked loop (DPLL) architecture is used to track the amplitude, frequency, and phase of the input signal and to acquire the phase signal readout [6]. A digital

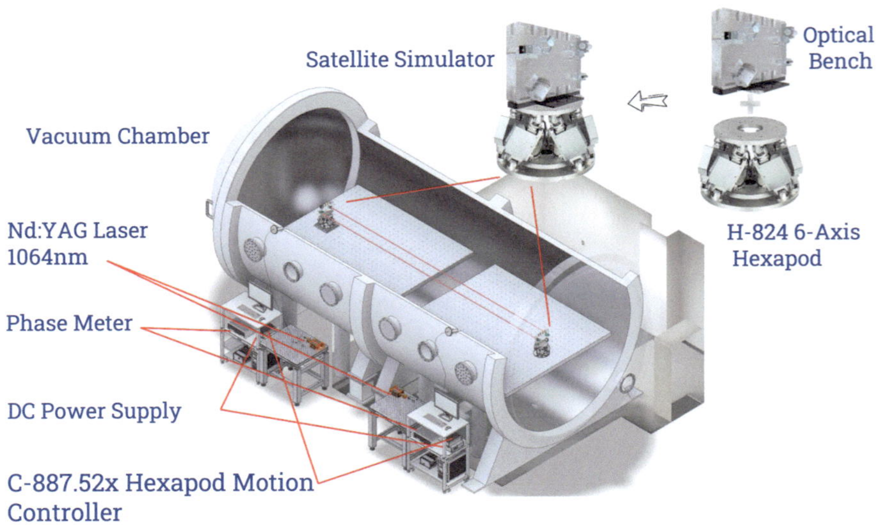

Fig. 22.3 Schematic diagram of the heterodyne interferometer experimental set-up in a vacuum chamber with 6-m vacuum chamber

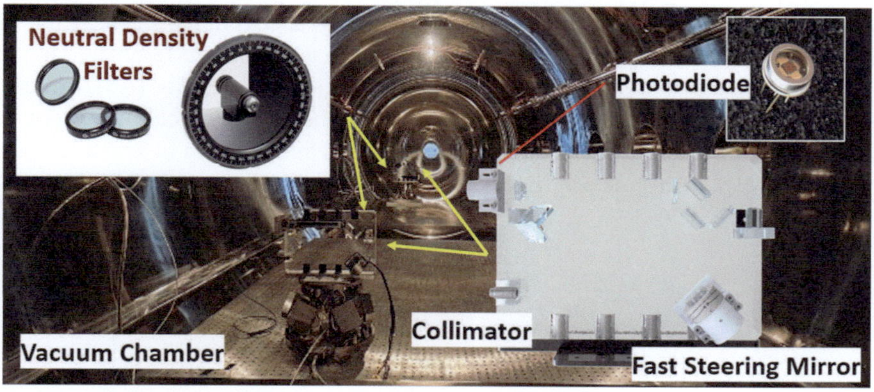

Fig. 22.4 Schematic diagram of the heterodyne interferometer experimental set-up in a vacuum chamber. The main interferometer, consisting of the rotation tables and optical bench, is placed inside the vacuum chamber

Table 22.1 Parameter configuration for the experimental setup

Description	Symbol	Value	Unit
Laser wavelength	λ	1064	nm
LO beam power	P_{LO}	0.2	mW
TX beam power	P_{TX}	19	mW
RX beam power	P_{RX}	0.2	nW
QPD slit width	δ	30	μm
QPD radius	r_0	1	mm
QPD responsivity	η	0.6	/
Waist radius of Gaussian Beam	ω_0	1.5	mm
Receive aperture radius	r_a	2	mm
Magnification factor of imaging system	M	5:1	/

phase-locked loop(DPLL) can also be employed to precisely maintain a frequency difference of 10 MHz in heterodyne interferometer. The hardware implementation encompasses both phase readout via digital PLL and phase locking for extremely weak optical power. The relevant parameters of this experimental setup are presented in Table 22.1.

22.4 Results

To better understand the measurement sensitivity of the laser interferometer prototype, we maintained stable environmental conditions and measured the drift of the interferometer in the vacuum environment. Then, the measuring data are analyzed with amplitude spectrum (AS). The linear amplitude spectrum density (LASD) of the displacement measurement signal obtained from the laser interferometer, in the absence of an external signal input. The amplitude spectral density (ASD) results are shown in Fig. 22.5. Laser ranging system obtained a theoretical measurement noise of 10 $\mathrm{nm}/\sqrt{\mathrm{Hz}}$ at 0.1 Hz that could satisfy the requirement of scientific missions. Figure 22.6 illustrates the linear spectral densities of the measured Differential

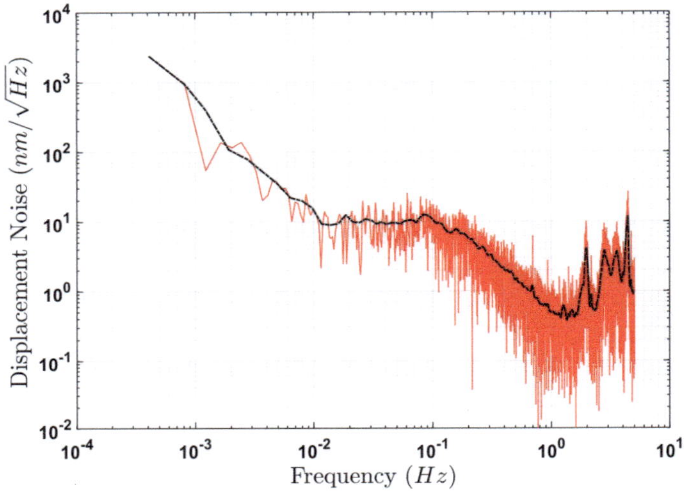

Fig. 22.5 Linear spectral densities of displacement noise

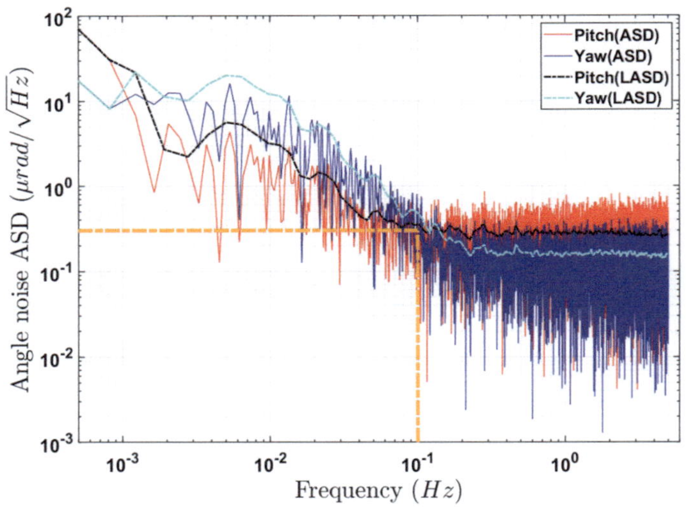

Fig. 22.6 Linear spectral densities of Angle measurement noise by Differential Wavefront Sensing(DWS)

Wavefront Sensing (DWS) signals for both horizontal and vertical directions under conditions. The angular measurement resolution exceeds 0.3 μrad/$\sqrt{\text{Hz}}$ at 0.1 Hz with differential wavefront sensing technique.

22.5 Conclution and Discussion

We set up an experimental facility for a ground-based demonstration of a transponder-typelaser interferometer prototype and establish a digital PLL to lock the phase of the slave laser to the phase of the master laser with the power of 0.2 nW. The DPLL is used to lock the frequency at 10 MHz, with a continuous phase lock time of at least one day, indicating that the device has good stability and can detect low-frequency signals. At present, the required measurement accuracy of this has been preliminarily achieved, and we expect to conduct numerous subsequent experimental tests. The resulting figure shows that the peak is clearly due to environmental disturbances when the frequency exceeds 1 Hz. In the future, we will focus on improving the laboratory environment to optimize the interferometer's performance.

References

1. Y. Hsien-Chi, Y. Qi-Zhong, L. Yu-Rong, W. Ying, L. Jun, Intersatellite laser ranging with homodyne optical phase locking for space advanced gravity measurements mission. Rev. Sci. Ins. **82**(4) (2011)
2. B.S. Sheard, G. Heinzel, K. Danzmann, D.A. Shaddock, W.M. Klipstein, W.M. Folkner, Intersatellite laser ranging instrument for the grace follow-on mission. J. Geodesy **86**(12), 1083–1095 (2012)
3. J. Luo, L. Sheng Chen, H. Zong Duan, Y. Gui Gong, S. Hu, J. Ji, Q. Liu, J. Mei, V. Milyukov, M. Sazhin, Tianqin: a space-borne gravitational wave detector. Classical Quant. Gravity **33**(3), 035010 (2015)
4. P. Amaro-Seoane, H. Audley, S. Babak, J. Baker, E. Barausse, P. Bender, E. Berti, P. Binetruy, M. Born, D. Bortoluzzi, et al., Laser interferometer space antenna. arXiv preprint arXiv: arXiv:1702.00786 (2017)
5. V. Anna-Maria A van, C. J. Killo,. Hydroxide catalysis bonding for astronomical instruments. Adv. Opt. Technol. **3**(3), 293–307 (2014)
6. G. Heinzel, M. Dovale Álvarez, A. Pizzella, N. Brause, J. José Esteban Delgado, Tracking length and differential-wavefront-sensing signals from quadrant photodiodes in heterodyne interferometers with digital phase-locked-loop readout. Phys. Rev. Appl. **14**(5), 054013 (2020)

Chapter 23
Decoupling Analysis of Multi-Degree-Of-Freedom Structural Stability Measurement for Gravitational Wave Detection Telescopes

Jian Luo, Bohong Li, Xinyu Li, and Yong Yan

Abstract The gravitational wave telescope, a critical payload of space-based gravitational wave detection satellites, experiences significant optical path stability issues due to deformations in its support structure. These deformations are primarily caused by thermal gradients and involve multiple degrees of freedom, underscoring the importance of studying methods to measure multi-degree-of-freedom deformations of the telescope frame. However, in the measurement process, the coupling of multiple degrees of freedom can adversely affect measurement accuracy, necessitating thorough decoupling analysis. Here, we introduce a preliminary technology for multi-DOF deformation measurements through decoupling analysis and noise evaluation of the single-arm measurement optical path. The measurement system meets the requirement of having background noise less than 10 pm/Hz$^{1/2}$ near 1 Hz and 2 nm/Hz$^{1/2}$ near 1 mHz. This analysis could help a structural deformation measurement framework to establish and provide broader insight into the response of methodological system for telescopes.

23.1 Introduction

Gravitational waves are ripples in spacetime caused by the acceleration of massive objects, and their detection relies on space-based telescopes, which serve as critical payloads for beam expansion and contraction. This analysis aimed to deepen the understanding of structural stability, which is inherently linked to the optical path stability—a critical performance indicator for these telescopes. [1–3]. There has been limited research on the structural stability of telescopes, and studies on the

J. Luo · B. Li · X. Li · Y. Yan (✉)
MOE Key Laboratory of TianQin Mission, TianQin Research Center for Gravitational Physics and School of Physics and Astronomy, Frontiers Science Center for TianQin, Gravitational Wave Research Center of CNSA, Sun Yat-Sen University (Zhuhai Campus), Zhuhai, China
e-mail: yanyong5@mail.sysu.edu.cn

J. Luo et al. (eds.), *Proceedings of the 7th International Workshop on the TianQin Science Mission; 25–26 April 2024, Hong Kong SAR, China*, Springer Proceedings in Physics 428, https://doi.org/10.1007/978-981-96-7803-7_23

decoupling of multi-degree-of-freedom deformations are lacking. The aim of this study is to analyze and validate the feasibility of a multi-degree-of-freedom structural stability measurement method for the telescope mount. The solution involves noise analysis and decoupling studies of the measurement system, with preliminary validation conducted through simulations and experiments. This study introduces a triaxial deformation measurement technology specifically designed for the optomechanical configuration of the TianQin telescope, providing a critical contribution to ensuring its stability and performance in gravitational wave detection.

23.2 Multi-Degree-Of-Freedom Deformation Measurement Technology

The displacement measuring interferometer (DMI) system of our apparatus consists of highly stable interferometers, reflector mirrors, fiber-optic pickups and supporting electronics. Lasers and acousto-optic modulators positioned outside the vacuum chamber generate heterodyne lasers that pass through the optical windows of the vacuum tank into the test chamber. Once outside the vacuum tank, the laser is distributed across an integrated optical device platform, splitting into several measurement pathways before entering the experimental chamber through the optical windows. In the telescope frame, specially installed test mirrors reflect these measurement lasers toward photodetectors located inside the test chamber. Data acquisition cards then collect this reflected laser data, extracting phase information to determine the deformations of the telescope frame. The test chamber is vacuum-sealed and maintains a constant temperature, isolated from the experimental chamber to ensure optimal interferometry conditions. The external chamber is thermally isolated using a heat sink to mitigate environmental thermal disturbances, while its interior walls are equipped with heating coils and radiation lamps to mimic the space operating conditions of the telescope (Fig. 23.1).

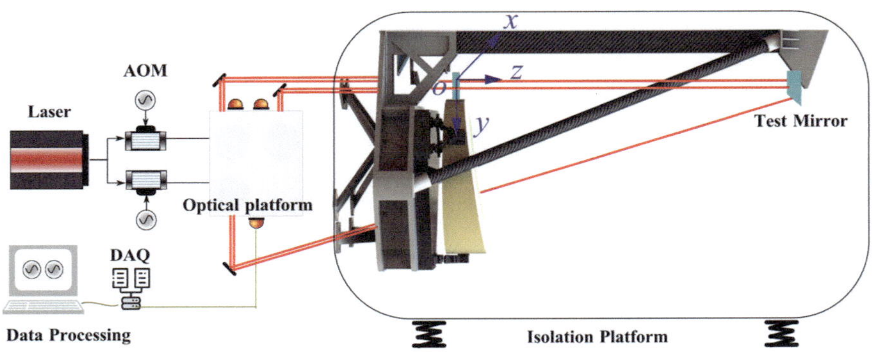

Fig. 23.1 Schematic of 3DOF deformation measurement system

Different mirrors experiencing the same rigid-body movements affect the optical path differently, with primary and secondary mirrors having a more significant impact than tertiary and quaternary mirrors. The influence of mirror displacements on the optical path varies with the degree of freedom; translations in the DZ (Along the optical axis of the primary mirror) and DY (Perpendicular to the square support rod, towards the primary mirror) directions and rotations around the RX (Rotation around the x-axis) axis are particularly significant. Accordingly, this study measures three degrees of freedom: translations in the z and y directions and rotation around the RX axis between the primary and secondary mirrors. Tracking changes at these characteristic points provides measurements of deformations along these sensitive directions of the telescope frame [6].

To measure the relative displacement between the primary and secondary mirrors, we designed an optical path based on common-mode suppression technology, as shown in Fig. 23.2. This optical path achieves extremely low background noise levels through noise cancellation. To mitigate the impact of thermal noise, a three-stage thermal insulation strategy was implemented. First, components with the highest heat generation, such as power supplies, were subjected to the first stage of insulation. Second, components with lower heat generation, such as lasers and AOMs, were placed under the second stage of insulation. Finally, optical components underwent the third stage of insulation. This insulation strategy effectively reduces heat transfer, significantly minimizing the impact of thermal flux on the optical path. Additionally, the optical components were integrated onto the optical platform using an adhesive bonding method, which lowers the temperature-displacement coupling coefficient of the measurement system, further reducing the influence of thermal noise.

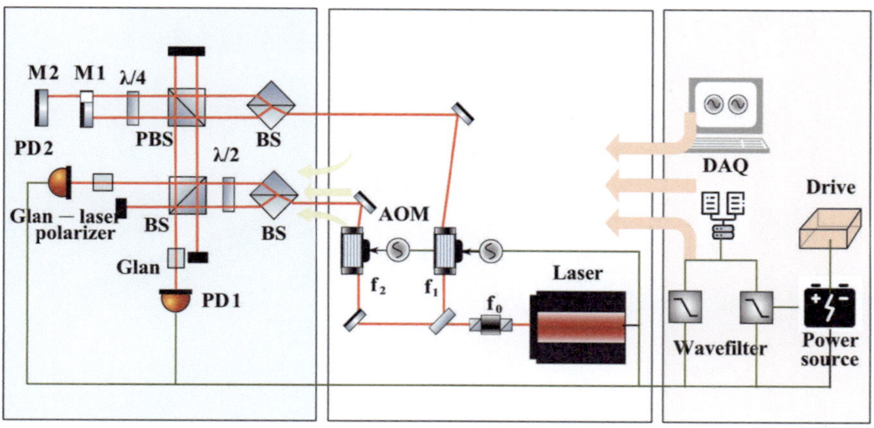

Fig. 23.2 Schematic of single-link measurement optical path design

23.3 Analysis and Simulation

Geometric coupling refers to the dependency relationships among displacements across different degrees of freedom, which are determined by the shape and size of the object. When conducting multi-DOF displacement measurements, geometric coupling significantly impacts the precision and accuracy of the measurements. To effectively address this issue, We use mathematical modeling and decoupling analysis to establish an accurate displacement calculation model, as illustrated in Fig. 23.2.The response matrix derived from the aforementioned method for three-degree-of-freedom displacement and each optical sensor pathway is:

$$\begin{pmatrix} \mathrm{d}I1 \\ \mathrm{d}I2 \\ \mathrm{d}I3 \end{pmatrix} = \begin{pmatrix} 1 & 0 & 0 \\ 1 & 0 & b \\ \cos\alpha & -\sin\alpha & (b+d-n)\cos\alpha + \frac{\mathrm{n}}{\cos\alpha} \end{pmatrix} \begin{pmatrix} \mathrm{d}DZ \\ \mathrm{d}DZ \\ \mathrm{d}RX \end{pmatrix} \tag{23.1}$$

where dI1, $dI2$ and $dI3$ are the displacement values measured by three laser measurement interferometric pathways, DY and DZ are the displacements of the test mirror along the y-axis and z-axis, respectively, RX is the rotation about the x-axis, b is the distance between the optical axes of two laser pathways, d is the distance between the projections of the laser optical axes on the plane of the secondary mirror's test mirror, n is the distance from the optical axis of $I3$ to the edge of the secondary mirror's test mirror, and α is the size of the dihedral angle of the test mirror (Fig. 23.3).

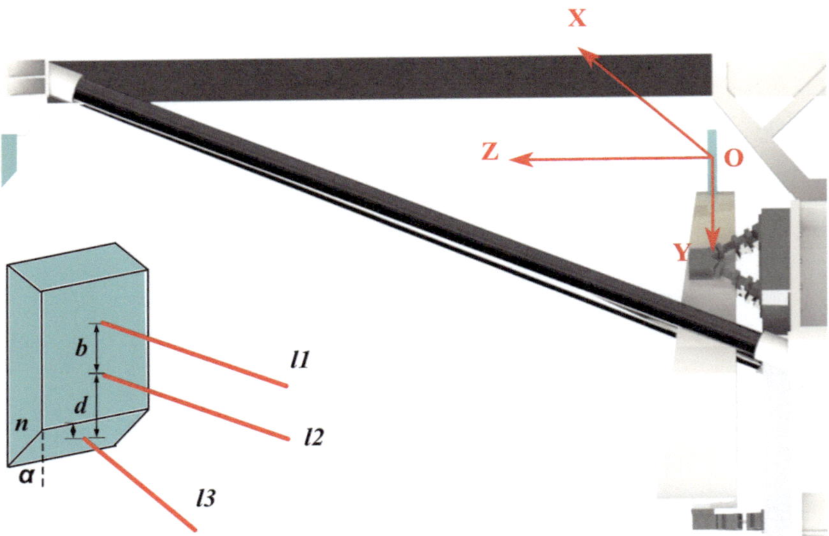

Fig. 23.3 Schematic diagram of coordinate system and response matrix parameters

In the Zemax optical simulation software, a simulation analysis of the optical path was conducted under non-sequential mode. A model of the 3DOF measurement optical path was established to match the actual measurement pathways, as illustrated. The study involves altering the relative positions of the primary and secondary mirrors to induce three degrees of rigid-body displacement in each. Subsequently, we compare the displacement values calculated by different detectors when the secondary mirror is at its original position versus after it has undergone three degrees of rigid-body displacement. The displacement values derived from the detector's phase measurements aligned with the changes in the secondary mirror's position validate the previously established coupling matrix relationships. This preliminary validation confirms the accuracy of the coupling matrix, providing a basis for experimental guidelines.

23.4 Experiment and Results

Currently, In order to meet the measurement requirements of the telescope prototype, the telescope frame deformation measurement system is designed to achieve a background noise level less than 2 nm/Hz$^{1/2}$ at 1 mHz. The noise specifications are distributed as shown in Table 23.1. To validate the feasibility of the optical path design in the measurement scheme, this study conducts a theoretical analysis of the allocated noise metrics followed by experimental measurements.

The experimental device of the single-arm interferometer is shown in Fig. 4a, This section presents preliminary experimental validations of the measurement scheme designed in this study. The noise levels of the single-arm measurement optical path, as depicted in Fig. 4b, show that the single-path interferometer exhibits a displacement noise floor of 10 pm/Hz$^{1/2}$ near 1 Hz and 2 nm/Hz$^{1/2}$ near 1 mHz. The noise levels of the measurement system in the low-frequency range below 1 mHz are predominantly influenced by environmental temperature noise. In contrast, at higher frequencies above 10 mHz, the system's noise is mainly constrained by laser frequency noise, inherent noise in phase acquisition, and vibrational disturbances.

23.5 Prototype Telescope

We are currently testing a prototype of the telescope design to allow us to validate the multi-degree-of-freedom structural stability measurement model. Figure 5a shows a photograph of the telescope. The telescope prototype was fabricated using conventional materials, and the mirrors have not yet been installed. Figure 5b shows the integrated optical component platform. An integrated platform can perform measurements for a single degree of freedom.

Table 23.1 Allocation and requirements of optical path noise indicators for tele-scope deformation measurement system

Measuring system components	Main noise source	Noise allocation (nm/ $Hz^{1/2}$@1 mHz)	Experimental validation
Laser	Laser frequency noise	0.3	Laser frequency noise < 4.2×10^7 Hz/ $Hz^{1/2}$@1 mHz
	Laser Amplitude Noise	0.1	Laser amplitude noise < 60 dB@1 mHz
Laser modulation coupling noise	Acousto-optic modulator coupling noise	0.3	Equivalent displacement noise < 0.3 nm/ $Hz^{1/2}$@1 mHz
	Fiber -temperature coupling noise	0.2	Fiber temperature fluctuation $\delta T \leq 15$ mK/$Hz^{1/2}$@1 mHz
Optical device platform	Temperature displacement coupling noise	0.8	Temperature displacement coupling coefficient dS/dT = 50 nm/K, Test environment temperature fluctuation $\delta T \leq 1.5$ mK/ $Hz^{1/2}$@1 mHz
	Vibration displacement coupling noise	0.2	Vibration displacement coupling coefficient dS/ dx = 2×10^{-5} m/m, Test environment vibration fluctuation $\delta X \leq 1 \times 10^{-3}$ m/s / $Hz^{1/2}$@1 mHz
	Air refractive index noise	0.2	temperature fluctuations $\delta T \leq 15$ mK/ $Hz^{1/2}$@1 mHz
	Polarization noise	0.1	Equivalent displacement noise < 0.2 nm/ $Hz^{1/2}$@1 mHz
	Stray light noise	0.2	Stray light intensity is lower than -30 dB compared to the main beam
Signal acquisition And Data processing	Detector noise	0.1	Relative intensity noise lower than -100 dB/ Hz@1 mHz
	Phase read noise	0.01	Phase meter equivalent displacement noise floor less than0.01 nm/ $Hz^{1/2}$@1 mHz
The noise of the Measure system		≈1	≈2

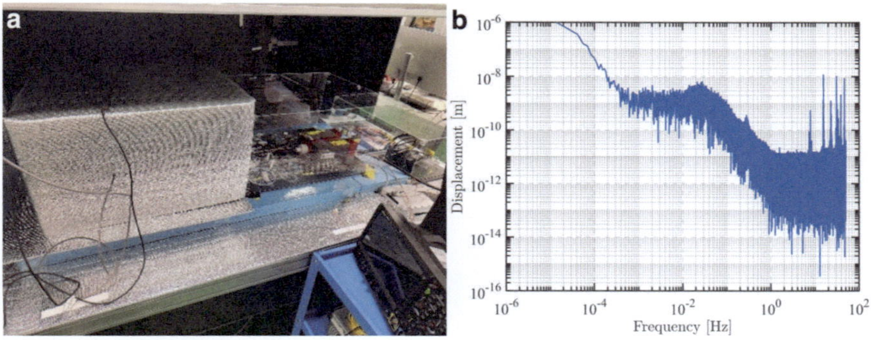

Fig. 23.4 Single-link interferometer experiment device and background noise. **a** Photo of device **b** Path-length measurement sensitivity of the device prototype

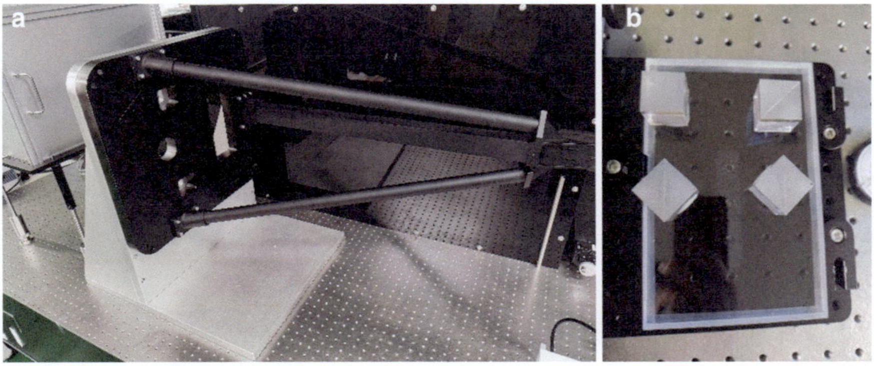

Fig. 23.5 Single-link interferometer experiment device and background noise. **a** Photo of device **b** optical component integration platform, multi-degree-of-freedom measurement probe

23.6 Conclusion

This paper has given an account of and the reason for the widespread use of high-precision measurement techniques for structural deformations in space-based gravitational wave detection telescopes, initially establishing a structural deformation measurement framework and methodological system for telescopes. The measurement system meets the requirement of having background noise less than 10 pm/$Hz^{1/2}$ near 1 Hz and 2 nm/$Hz^{1/2}$ near 1 mHz, preliminarily proving the feasibility of the multi-path heterodyne interferometry measurement method. The current study was unable to achieve the required stability with the telescope prototype due to the use of conventional materials, highlighting the need for developing more stable telescope structures. Additionally, the interferometer's precision in this study fell short of meeting the final requirements, indicating that future research should focus on further improving its accuracy.

References

1. M. Peterseim, D.I. Robertson, K. Danzmann et al., LISA interferometer sensitivity to spacecraft motion. Adv. Space Res. **25**, 1143–1147 (2000)
2. J. Luo, L.S. Chen, H.Z. Duan et al., TianQin:a space-borne gravitational wave detector. Class. QuantumGrav. **33**, 035010 (2016)
3. M. Yu, J.C. Li, H.G. Lin et al., Optical system design of large aperture space gravitational wave telescope. Opt. Eng. **62**, 065107 (2023)
4. L. Blecha, D. Zindel, H. Cottard et al., Analytical optimization and test validation of the submicron dimensional stability of the CHEOPS space telescope's CFRP structure. Proc. SPIE **9912**, 99121G (2016)
5. W.A. Klop, A.L. Verlaan, Dimensional stability testing in thermal vacuum of the CHEOPS optical telescope assembly. Proc. SPIE **9904**, 990437
6. J. Luo, J. Song, S.J. Fang et al., Decoupling study and noise analysis of multi-degree-of-freedom deformation measurement method for space gravitational wave detection telescope. Opto-Electron. Eng. **51**, 230211 (2024)

Chapter 24
Analysis of Material Selection for the TianQin Telescope

Zihan Wang, Ming Li, and Hongchao Zhao

Abstract TianQin is a space-based gravitational wave observatory in a high geocentric orbit, exposed to space radiation heat sources. This study evaluates the thermal properties combined with the physical and mechanical properties of several materials, concluding that ULE® and Zerodur® are superior choices for optical path stability within the 0.1 mHz to 1 Hz frequency range, with Zerodur showing superior performance. Their ultra-low coefficients of thermal expansion (CTE) ensure minimal length variations due to temperature fluctuations. Their high thermal conductivity and heat capacity further enhance their ability to reduce noise effectively.

24.1 Introduction

To directly detect gravitational wave signals within the 0.1 mHz to 1 Hz range, the TianQin project utilizes a Michelson principle [1, 2], which operates in a geocentric orbit approximately 10^5 km from Earth [3]. The system forms an equilateral triangle laser interferometer measurement system (IMS) formation with arm lengths of 170,000 km. Detecting gravitational wave-induced space distortions via changes in optical paths between test masses. Stringent noise limitations are essential with the weak nature of these signals. In IMS, the optical path noise from the telescope is critical, arising from positional variations between primary and secondary mirrors, affecting interferometric accuracy. In the in-orbit environment, there are significant temperature differences and fluctuations between the interior and exterior of the satellite. The cantilever between the primary and secondary mirrors of the telescope, acting as a structure bridging the internal and external environments, undergoes complex temperature variations under these conditions. We refer to these variations as temperature noise. Under the influence of temperature noise, the thermal expansion

Z. Wang (✉) · M. Li · H. Zhao
MOE Key Laboratory of TianQin Mission, TianQin Research Center for Gravitational Physics and School of Physics and Astronomy, Frontiers Science Center for TianQin, Gravitational Wave Research Center of CNSA, Sun Yat-Sen University (Zhuhai Campus), Zhuhai, China
e-mail: wangzh368@mail2.sysu.edu.cn

J. Luo et al. (eds.), *Proceedings of the 7th International Workshop on the TianQin Science Mission; 25–26 April 2024, Hong Kong SAR, China*, Springer Proceedings in Physics 428, https://doi.org/10.1007/978-981-96-7803-7_24

and contraction of the distance between the primary and secondary mirrors introduce optical path noise. Accordingly, this study analyzes the properties of several common aerospace materials and provides recommendations for material selection for the TianQin telescope.

24.2 Optical Design and Indicator Requirements

The function of the TianQin Space Telescope is to realise the signal transmission between the two spacecraft in the two places, to accurately measure the distance between the test mass blocks. In terms of optical design, the off-axis four-inverse focusless structure scheme (see Fig. 24.1) is adopted. The primary mirror (M1) is parabolic, the secondary mirror (M2) is hyperbolic, and the third and fourth mirrors (M3 and M4) are aspherical [4].

The current total noise budget for the optical path difference in a single-link measurement system is given. In the total optical path noise budget, the coupling noise introduced by changes in the distance between the primary and secondary mirrors of the telescope accounts for nearly half, which is given by Eq. (24.1), where $\mathbf{0.1}\boldsymbol{mHz} \leq f \leq \mathbf{1}\boldsymbol{Hz}, f_0 = \mathbf{7}\boldsymbol{mHz}$.

$$NB_{OP,SC} = 0.4\frac{\boldsymbol{pm}}{\sqrt{\boldsymbol{Hz}}} \bullet \sqrt{1 + (\frac{f_0}{f})^4} \tag{24.1}$$

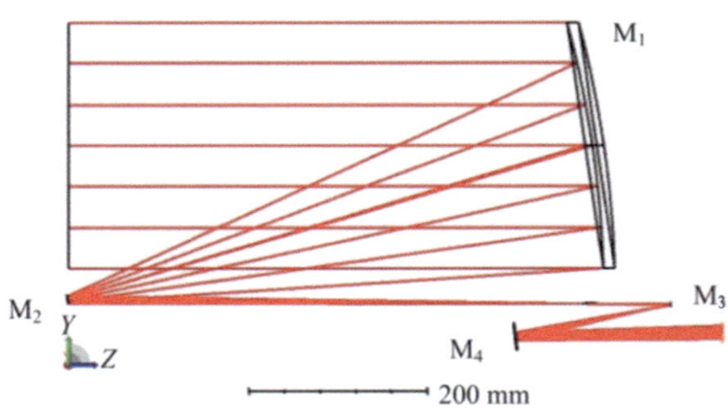

Fig. 24.1 Optical system of the TianQin telescope

Table 24.2 Main performance parameters

Materials	Density $\rho(g/cm^3)$	Elastic modulus $E(Gpa)$	Thermal conductivity $\lambda(W/(m \bullet K))$	Linear expansion coefficient $\alpha(10^{-6}/\mathrm{K})$	Specific heat capacity C $(10^2 J/(kg \bullet K))$
SiC	3.15	410	155	2.51	4.3
ULE	2.21	67	1.31	0.03	4.6
Zerodur	2.53	91	1.63	0.01	8.0
CFRP	1.80	95	6	0.15	1.0
Invar	8.05	200	10	2.11	2.3
Aluminum	2.68	78	167	23.61	8.9

24.3 Material Selection

24.3.1 Alternative Materials

In considering gravitational rebound deformation in the space microgravity environment, it is crucial to integrate both the thermal and mechanical properties of materials alongside their thermal characteristics. Materials with ultra-low coefficients of thermal expansion (CTE) to address challenges faced by high-precision space telescopes. Examples include ultra-low expansion glass (ULE) from Corning [5], Zerodur from SCHOTT [6], silicon carbide (SiC) [7, 8], and continuous fiber-reinforced polymeric composites (CFRP) [9, 10]. Table 24.2 provides an overview of commonly used materials in space telescopes.

24.3.2 Mechanical Properties

Due to the stringent requirements for optical path stability, aluminum and Invar materials show clear drawbacks compared to others. SiC offers high specific stiffness and excellent thermal stability, promoting efficient thermal equilibrium. Conversely, ULE and Zerodur boast significantly lower coefficients of thermal expansion (CTE), ensuring superior thermal stability but require optimized structures or additional support to mitigate vibration and stress during transportation due to their low specific stiffness and brittleness. Zerodur, with higher crystallization, offers enhanced mechanical strength over ULE. CFRP, valued for its high stiffness and low density, is prone to resin-based vacuum outgassing during transportation and in-orbit operations, potentially contaminating optical elements and introducing stray light noise [11]. Figure 24.2 presents a comprehensive evaluation of specific stiffness and thermal properties, facilitating comparative analysis of different materials.

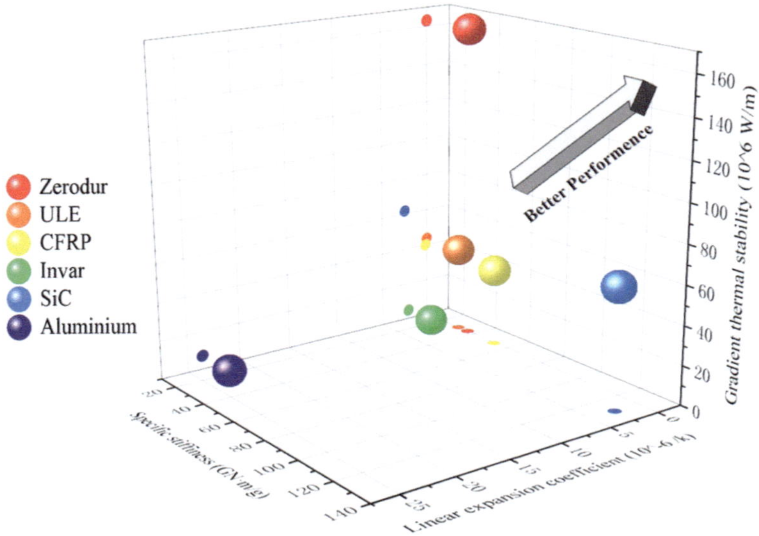

Fig. 24.2 Comparison of material properties. The direction indicated by the black arrow represents better performance

In conclusion, Zerodur, ULE, CFRP, and SiC, particularly Zerodur, show promise for constructing the supporting cantilever between the primary and secondary mirrors. Subsequent sections will delve into these materials, assessing their suitability for the thermal environment of the TianQin Space Telescope in orbit.

24.3.3 Thermal Properties

The TianQin telescope will be mounted on an annular structure using brackets at the primary mirror end. Previous studies indicate that this structure maintains a temperature of approximately 295 K, with a stability of $100\mu K/Hz^{1/2}$@0.1mHz ~ 1Hz. In contrast, the secondary mirror, situated closer to outer space and thermally regulated by a passive insulation cylinder, will be several degrees cooler at thermal equilibrium, depending on the cantilever material. The gravitational wave detector operates by analyzing laser phase shifts, necessitating high structural stability, predominantly influenced by temperature noise during in-orbit operations. This paper investigates the effects of minor thermal fluctuations under thermal equilibrium conditions. In orbit, the telescope's thermal stability is influenced solely by heat conduction and radiation, excluding convection.

After research, the in-orbit thermal environment is categorized into three primary thermal loads (see Fig. 3a): heat conduction from the satellite's internal structure (Q1), radiated heat from the insulation cylinder into the telescope cavity (Q2), and

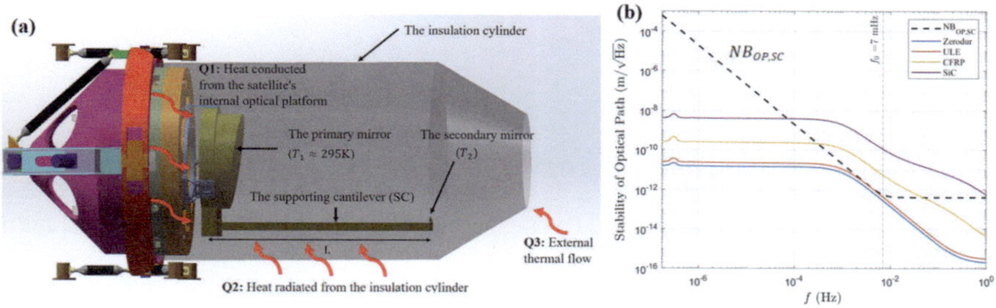

Fig. 24.3 **a** In-orbit thermal environment. **b** The optical path stability of four alternative materials

external thermal flow (Q3). These loads include both direct current (DC) and alternating current (AC) components; however, this study focuses exclusively on the AC components to examine relative temperature changes.

Here is the conclusion: Curves depicting the optical path stability of cantilevers made from various materials within the 0.1mHz to 1Hz target frequency range have been plotted (see Fig. 3b). The black dashed line represents the noise budget (Eq. (24.1)). Among the four alternative materials, only the curves for ULE and Zerodur meet the noise budget requirements, with Zerodur exhibiting lower noise levels compared to the others.

24.4 Conclusion

Based on the evaluation of thermal and mechanical properties, ULE, and Zerodur emerge as superior choices among the four alternative materials considered. Within the 0.1mHz to 1Hz frequency range, these materials demonstrate curves that align with the noise budget requirements, with Zerodur specifically showing lower noise levels compared to SiC and CFRP. The favorable performance of ULE and Zerodur can be attributed to their ultra-low coefficients of thermal expansion (CTE), which minimize variations in cantilever length due to temperature fluctuations. Additionally, their thermal properties, including thermal conductivity and heat capacity, contribute to their effective noise reduction capabilities in high-precision applications such as space telescopes. This evaluation underscores the critical role of material selection in achieving stable and reliable optical systems for precise astronomical observations in demanding environmental conditions.

References

1. J. Luo et al., TianQin: a space-borne gravitational wave detector. Class. Quantum Gravity **33**, 035010 (2016)
2. J. Mei, et al, The TianQin project: current progress on science and technology. Prog. Theor. Exp. Phys. 05A107 (2021)
3. B.B. Ye, X. Zhang, M.Y. Zhou et al., Optimizing orbits for TianQin. Int. J. Mod. Phys. D **28**(9), 1950121 (2019)
4. Z. Fan, L. Zhao, J. Peng, et al, Recommended high performance telescope system design for the TianQin project. Physics.Optics. (2022)
5. S. Fan, M. Yu, D. Qing et al., Application and development status of ultra-low expansion microcrystalline glass (in Chinese). AcIOL **41**(S1), 210–214 (2012)
6. Forty years of ZERODUR® mirror substrates for astronomy–Review and Outlook
7. E.V Seina, Y. TOULEMONTa, F. Safaa, M. DURANa, P. DENYb, D.D CHAMBUREc, et al, A Phi 3.5 m SiC telescope for HERSCHEL Mission.Proc. SPIE. **4850**, 606–618 (2002)
8. K. Hidehiro, O. Takashi, N. Takao et al., Cryogenic optical performance of the ASTRO-F SiC telescope. Appl. Opt. **44**(32), 6823–6832 (2005)
9. N. Desnoyers, M.-A. Boucher, P. Goyette, Design guidelines for high dimensional stability of CFRP optical bench. Optomech. Eng. **8836**, 883609 (2013)
10. W. Li, H. Liu, Structural stability of carbon fiber composite precision support components in space optical remote sensors (in Chinese). Opt. Precis. Eng. **11**, 2173–2179 (2018)
11. T. Kamiya, T. Mizutani, Studies on hygroscopic behavior of composite materials for highly stable satellite structures. J. Soc. Mater. Sci. Japan. **66**(4), 275–281 (2017)

Chapter 25
Raw Data Processing and Analysis of Inter-Satellite Laser Interferometric Ranging

Heng Yin, Yihao Yan, Min Zhong, Zitong Zhu, Changqing Wang, and Wei Feng

Abstract The high-precision inter-satellite laser interferometer is the key payload of the TianQin mission satellite, and the raw data processing of laser interferometric measurements is a crucial step in achieving its scientific objectives. Based on the measurements from the GRACE-FO gravity satellite, this study investigated the characteristics of laser ranging interferometer (LRI) measurements and raw data processing procedures. To validate the reliability of the data processing techniques, this study compares the processing results with RL04 product form NASA JPL and V50 product form AEI in terms of phase jump processing, scale factor estimation, biased range, and gravity field recovery results. The results show that the LRI product processed in this study demonstrates a precision of 1 nm/s/$\sqrt{}$Hz at 0.1 Hz, which is

H. Yin (✉)
MOE Key Laboratory of TianQin Mission, TianQin Research Center for Gravitational Physics and School of Physics and Astronomy, Frontiers Science Center for TianQin, Gravitational Wave Research Center of CNSA, Sun Yat-Sen University (Zhuhai Campus), Zhuhai, China
e-mail: yinh39@mail2.sysu.edu.cn

Y. Yan
Max-Planck-Institut Für Gravitationsphysik, Albert-Einstein-Institut) and Institut Für Gravitationsphysik, Leibniz Universität Hannover, Hannover, Germany
e-mail: yihao.yan@aei.mpg.de

M. Zhong · W. Feng
School of Geospatial Engineering and Science, Sun Yat-Sen University, Guangdong Zhuhai, China
e-mail: zhongm63@mail.sysu.edu.cn

W. Feng
e-mail: fengwei@mail.sysu.edu.cn

Z. Zhu · C. Wang
State Key Laboratory of Geodesy and Earth's Dynamics, Innovation Academy for Precision Measurement Science and Technology, Chinese Academy of Sciences, Wuhan, China
e-mail: tong@apm.ac.cn

C. Wang
e-mail: whiggsdkd@apm.ac.cn

J. Luo et al. (eds.), *Proceedings of the 7th International Workshop on the TianQin Science Mission; 25–26 April 2024, Hong Kong SAR, China*, Springer Proceedings in Physics 428, https://doi.org/10.1007/978-981-96-7803-7_25

comparable to the V50 and better than RL04. Moreover, the geoid degree variance of gravity field based on the product in this study is superior to JPL's results for degrees above 30, and shows a good consistency with AEI's results. The LRI raw data processing algorithm and software shown in this study can effectively process the raw LRI measurements, providing a foundation for the future LRI data processing work of TianQin 2, while also offering experience for TianQin 3.

25.1 Introduction

The TianQin project, proposed by Sun Yat-sen University in 2014, is a space-based gravitational wave detection observatory [4]. By around 2035, it aims to deploy three identical satellites in a nearly equilateral triangular formation with arm lengths approximately 170,000 km at an altitude of about 100,000 km to detect gravitational waves within the frequency range of 0.1 mHz to 0.1 Hz. The TianQin project has outlined a "0123" technical roadmap [2]:

Step 0: Construct a ground-based laser ranging station and conduct lunar laser ranging experiments to acquire ranging and orbit determination capabilities for the TianQin satellites.

Step 1: Launch a single satellite to verify key technologies such as high-precision inertial sensing, micro-Newton thruster and drag-free control.

Step 2: Launch a pair of satellites for in-orbit verification of inter-satellite laser interferometry technology.

Step 3: Deploy the TianQin satellites for gravitational wave detection.

Currently, the tasks of step 0 and step 1 have been successfully completed with excellent results [3]. The TianQin 2 satellite for step 2 has been successfully approved and is expected to be launched before 2026. This mission will not only serve as a verification satellite for laser interferometry technology but also has special value for high-precision Earth gravity field detection. The inter-satellite laser interferometer is one of the core payloads for the TianQin 2 and TianQin 3 satellites, and the raw data processing is a critical step in achieving the mission's scientific objectives. The Gravity Recovery and Climate Experiment-Follow On (GRACE-FO) satellites carry the first laser ranging interferometer (LRI), which is of significant value for analyzing LRI measurements characteristics [1]. The Science Data System (SDS) is responsible for the ground data processing of GRACE-FO. The processing of payload data from Level-0 to Level-1B is completed by the Jet Propulsion Laboratory (JPL), which has released the official version of the LRI1B product: RL04 [8]. The Albert Einstein Institute (AEI) team, the primary developer of the GRACE-FO LRI system, has also conducted in-depth research on LRI raw data processing techniques and has released a series of LRI1B products, including versions V50 to V54 [5, 7]. For the purpose of developing key techniques of LRI data processing for TianQin 2 and TianQin 3, this study investigated the LRI raw data processing algorithm based on the GRACE-FO LRI data and derived the corresponding results.

25.2 Method

Figure 25.1 illustrates the GRACE-FO LRI raw data processing flow from LRI1A to LRI1B in this study. It should be noted that all the results in this study is derived from the method developed in [9] and its datasets (LRI1B SYSU S10) published in website: https://zenodo.org/records/13743822. The processed LRI1B data includes biased range and its first and second derivatives, as well as light time correction and its first and second derivatives. This paper focuses on the phase jump processing results and scale factor estimation results. Below is a brief introduction to the methods used in this paper. For detailed explanations, please refer to literatures: [7, 9, 5].

(1) Phase Unwrapping: This step involves converting the folded LRI measurements of four channels of quadrant photoreceiver (QPR) to unwrapped phase measurements, converting the unit from counts to cycles and averaging the four channel LRI measurements.
(2) Quality Detection: Detects data gaps and anomalies in LRI1A phase measurements and classifies and processes them accordingly.
(3) Time-tag Correction: converts the local time-tags of LRI phase measurements into the GNSS time-tags.
(4) Deglitching: Combines phase jump template methods with phase smoothing to eliminate phase jumps and anomalies.
(5) Merging phase by Transponder - Master and converting it to range.
(6) Light time correction: Corrects the biased range calculated from LRI phase measurements to instantaneous range between the satellites.

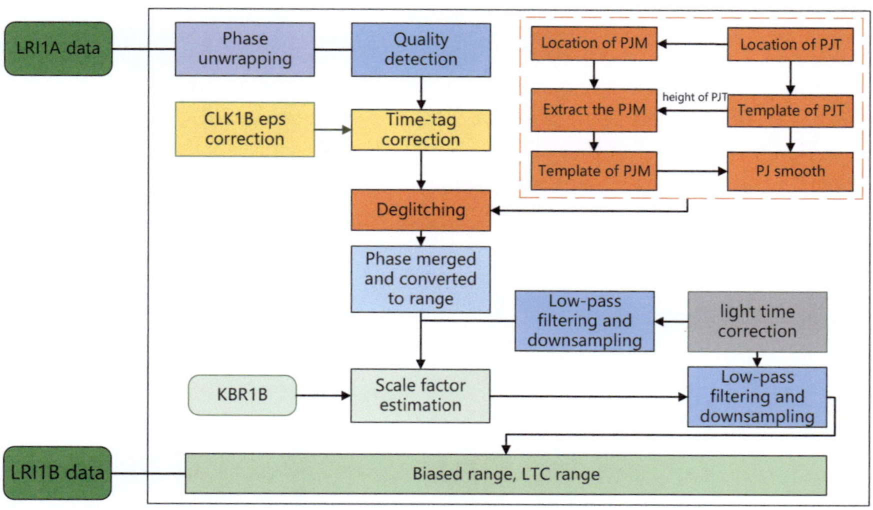

Fig. 25.1 LRI1A to LRI1B raw data processing flow chart

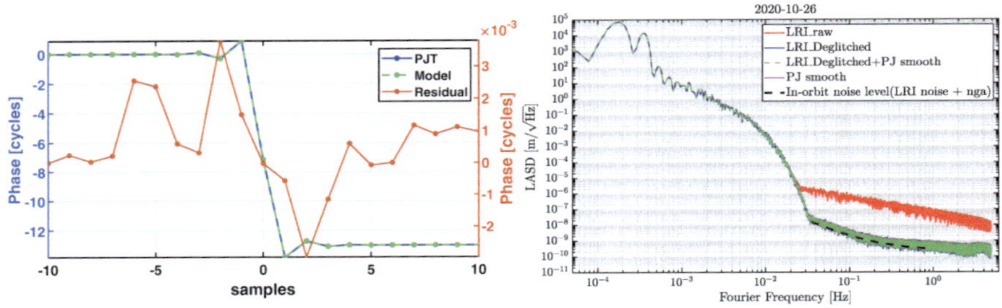

Fig. 25.2 The phase jump in time domain(left) and LRI spectrum of phase jump elimination result using template method(blue), and template + smoothing method(green) on October 26, 2020(right)

(7) Scale factor estimation: Uses the least squares method to estimate the scale factor between LRI and KBR ranging measurements and the time offset between LRI and KBR systems, correcting the range system error caused by laser frequency drift.

(8) Down sampling: To reduce data computation, LRI data are down sampled from the original 9.664 Hz sampling rate to 0.5 Hz. A CRN low-pass filter is applied before down sampling to prevent aliasing, as designed and parameterized according to Thomas (1999) and [7].

25.3 Results

25.3.1 Deglitching

The phase jumps (PJ) are abnormal step signals appearing in LRI phase measurements. The phase jump templates and phase smoothing are adopted to process it [9, 5, 7]. The left panel of Fig. 25.2 shows the PJ processing result with phase jump template method, with residual errors below 0.005 cycles. The right panel depicts the LRI range processing results in frequency-domain, achieving precision near 300 pm/$\sqrt{}$Hz@1 Hz.

25.3.2 Scale Factor Estimation

Due to the frequency drift of about 600 MHz of the LRI laser with respect to the nominal frequency, the scale factor of LRI is estimated using the cross-validation method with KBR/LRI intersatellite range rate [6]. Figure 25.3 shows the comparison of scale factor results between SYSU S10, AEI V50, and JPL RL04. Overall, there is good consistency among the scale factor estimation results. However, some outliers

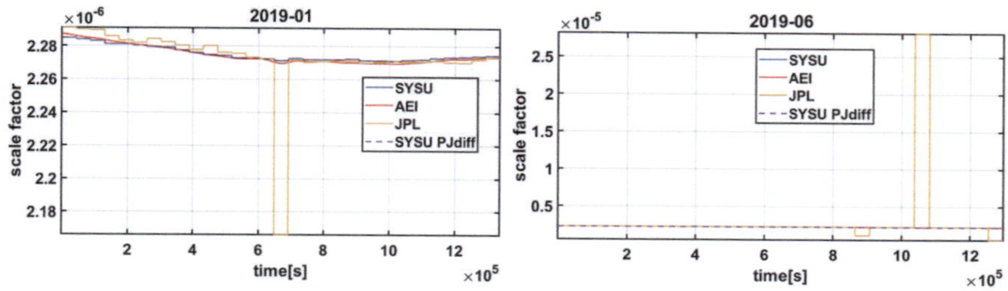

Fig. 25.3 Comparison of scale factor estimation results from SYSU, AEI, and JPL

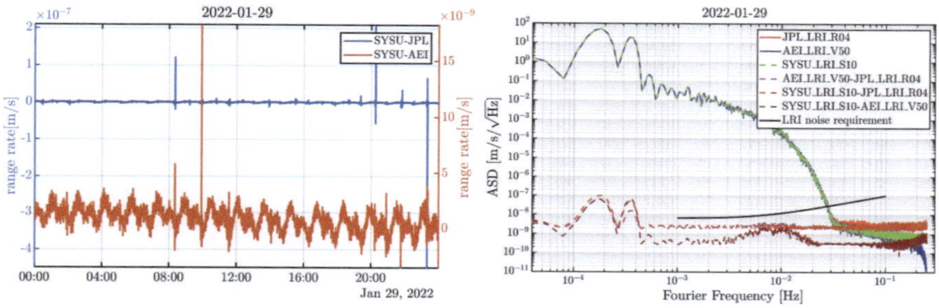

Fig. 25.4 Time–frequency comparison of LRI1B biased range rate from SYSU, JPL, and AEI

are present in the scale factor estimated by JPL, due to unprocessed Mega PJ and other anomalies in the JPL LRI results.

25.3.3 Biased Range Rate Comparison

Figure 25.4 compares the biased range rate of LRI1B between SYSU, JPL, and AEI. In the frequency band above 30 mHz, the products from all three institutions are below the LRI design noise level. S10 and V50 show the precision of 1 nm/s/$\sqrt{}$Hz at 0.1 Hz, which surpasses JPL's results. The differences observed at 1CPR and 2CPR are mainly caused by differences of the scale factor estimation results.

25.3.4 Gravity Field Results Comparison

Figure 25.5 compares the geoid degree variance among SYSU, AEI, and JPL. The gravity recovery method can be referenced in [10]. The results of SYSU and AEI demonstrate a good consistency and are superior to JPL's results for degrees beyond 30. This suggests higher noise levels in JPL's product at high degree, which may be

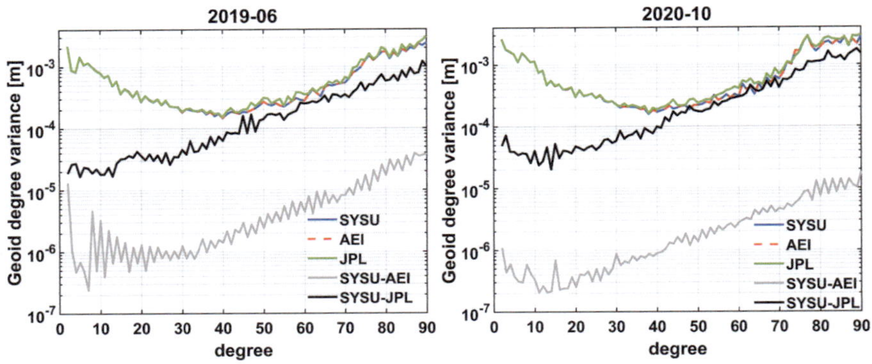

Fig. 25.5 Comparison of geoid degree variance of gravity field models inverted from LRI1B by SYSU, JPL, and AEI respectively

caused by the imperfect processing of phase jump and the abnormal value in scale factor.

25.4 Conclusions

Based on the GRACE-FO LRI measurements and the raw data processing algorithms from [9], this paper thoroughly shows the raw data processing results from LRI1A to LRI1B. For phase anomalies such as phase jumps and cycle slips, the phase jump template method combined with phase smoothing was employed, keeping the residuals after phase jump removal generally within 0.005 cycles. To address system errors caused by the absolute frequency drift of the on-orbit laser, the least squares method was used to accurately estimate the scale factor between the LRI and KBR range rates, mitigating the impact of the system errors. The precision of the S10 product is comparable to AEI products, with biased range rate precision reaching $1\text{nm/s}/\sqrt{\text{Hz}}$ at 0.1Hz, surpassing JPL's results, which may still have some deficiencies in their deglitching algorithm. In terms of gravity field recovery results, S10 has a good consistency with V50 and outperforms RL04 above degree 30.

The results demonstrate that the LRI raw data processing method shown in this paper provides a technical foundation for future TianQin 2 and also provide valuable reference for TianQin 3.

References

1. K. Abich, In-orbit performance of the GRACE follow-on laser ranging interferometer. Phys. Rev. Lett. **123**(3), 031101 (2019)

2. Y. Gong, Concepts and status of Chinese space gravitational wave detection projects. Nat. Astron. **5**(9), 881–889 (2021)
3. J. Luo, The first round result from the TianQin-1 satellite. Classi Quantum Gravity **37**(18), 185013 (2020)
4. J. Luo, TianQin: a space-borne gravitational wave detector. Class. Quantum Gravity **33**(3), 035010 (2016)
5. M. Misfeldt, Data processing and investigation for the GRACE-FO laser ranging interferometer. Master thesis, Leibniz Universität Hannover, Hannover, Germany (2019)
6. M. Misfeldt, Scale factor determination for the GRACE Follow-On laser ranging interferometer including thermal coupling. Remote. Sens. **15**(3), 570 (2023)
7. L. Müller, Generation of Level 1 data products and validating the correctness of currently available release 04 data for the GRACE follow-on laser ranging interferometer. Master thesis, Leibniz Universität Hannover, Hannover, Germany (2021)
8. H. Wen, Gravity recovery and climate experiment (GRACE) Follow-On (GRACE-FO) level-1 data product user handbook (2019)
9. H. Yin, Raw data processing of laser ranging interferometer based on GRACE-FO gravity satellite. Chin. J. Geophys.-Chin. Ed. (2024)
10. Z. Zhu, Comparative analysis of gravity field models of microwave and laser ranging data based on GRACE-FO. J. Geod. Geodyn. **42**(12), 1262–1268 (2022)

Chapter 26
A Thermal Deformation Optimization Method for CFRP Truss Structure of Gravitational Wave Telescopes

Bohong Li, Jian Luo, Sijun Fang, and Yong Yan

Abstract The gravitational wave detection telescope has stringent thermal deformation requirements that are difficult to meet with conventional aerospace materials. Carbon fiber-reinforced polymer (CFRP), with their excellent designability, allow for adjustments in the material's coefficient of thermal expansion (CTE) through layup design, making them a promising candidate material for the TQ telescope truss structure. This study designs the TQ telescope prototype by combining its truss structure with the customizable properties of CFRP. The method addresses excessive thermal deformation in the dz (separation) and dy (eccentricity) directions of M1 and M2. Considering the anisotropic characteristics of CFRP, the NSGA-II algorithm is proposed to solve the multi-objective layup design optimization problem. The optimization results are then incorporated into finite element simulations for in-depth analysis. Results indicate that the optimized structure significantly reduces thermal deformation, confirming the method's effectiveness.

26.1 Introduction

The detection of gravitational waves in space is achieved through the use of inter-satellite laser interferometry, which monitors changes in distance between two test masses separated by tens of thousands of kilometers. The structural performance of the telescope, as part of the interferometric system, directly affects the detection results. Among the various factors, the environmental temperature change is the main cause of structural deformation. The largest component inside the telescope is the support structure connecting the primary and secondary mirrors. This structure, which serves as a platform for supporting and fixing parts, significantly reflects

B. Li · J. Luo · S. Fang · Y. Yan (✉)
MOE Key Laboratory of TianQin Mission, TianQin Research Center for Gravitational Physics & School of Physics and Astronomy, Frontiers Science Center for TianQin, Gravitational Wave Research Center of CNSA, Sun Yat-Sen University (Zhuhai Campus), Zhuhai, China
e-mail: yanyong5@mail.sysu.edu.cn

J. Luo et al. (eds.), *Proceedings of the 7th International Workshop on the TianQin Science Mission; 25–26 April 2024, Hong Kong SAR, China*, Springer Proceedings in Physics 428, https://doi.org/10.1007/978-981-96-7803-7_26

the overall thermal deformation behavior of the telescope. Consequently, it requires in-depth analysis and optimization.

In the telescope system, the relative positional changes between the primary and secondary mirrors in different directions affect the optical path in disparate ways. The changes in separation (dz) and eccentricity (dy) directions have the most significant impact on the optical path [1, 2]. Based on the current environmental thermal analysis and optical tolerance allocation, the structural thermal deformation requirements for the relative position of the secondary mirror with respect to the primary mirror are as follows: dz < 0.55 μm/K and dy < 0.2 μm/K.

Three materials can meet the above thermal deformation requirements: Zerodur, Invar, and CFRP. Invar is not suitable for large telescope support structures due to its high density and magnetic properties. Therefore, there are two design approaches for the telescope's truss support structure [3]. One approach employs Zerodur as the primary material, relying on adhesive bonding or monolithic manufacturing techniques for component production. Due to low CTE (less than 2×10^{-8}/K) of Zerodur, this structure can significantly reduce thermal deformation [4]. However, this method demands high precision in bonding and manufacturing processes.

The second approach uses CFRP as the truss material, employing a segmented assembly design to simplify machining, assembly, and adjustment. Given that CFRP is a designable anisotropic material, this property can be leveraged to provide targeted structural design solutions that reduce dz and dy [5–7].

This study proposes an assembly telescope based on the characteristics of CFRP materials to reduce thermal deformation between the primary and secondary mirrors. While ensuring that the assembly structure meets requirements, the design combines material designability with the characteristics of truss structures. It introduces differentiated CFRP layup designs for truss members at different positions to minimize structural thermal deformation and meet design requirements.

26.2 Method

26.2.1 Truss Structure Thermal Deformation Model

Given the relatively mature state of the process of forming CFRP tubes, the structure of the telescope prototype is designed as a truss style. CFRP is selected from the M40J customized fibers provided by Changchun Long Aerospace Composite Materials Co., Ltd. The connecting components utilize Invar alloy of grade S4J32B, while the mirror body is constructed from Zerodur material, produced by Schott. The data for each material is presented in Table 26.1.

o investigate the relationship between the thermal deformation performance of the structure and the thermal expansion coefficient (CTE) of the truss structure, a thermal deformation model of the telescope truss structure is established.

Table 26.1 Candidate materials for truss support structure of telescope prototype

Material	Specification	Density	Engineering constants	CTE
Zerodur	ZERODUR® Expansion Class 1	$\rho = 2530kg/m^3$	E = 91GPa, $\nu =$ 0.24	$\alpha = 0.05 \times 10^{-6}/K$
Invar	S-4J32/B	$\rho = 7980kg/m^3$	E = 168GPa, $\nu =$ 0.28	$\alpha = 1 \times 10^{-6}/K$

As shown in Fig. 26.1, point U is taken as the observation point, representing the thermal deformation situation of the truss structure.

As shown in Fig. 26.2, there is a geometric relationship during thermal deformation of the telescope truss structure:

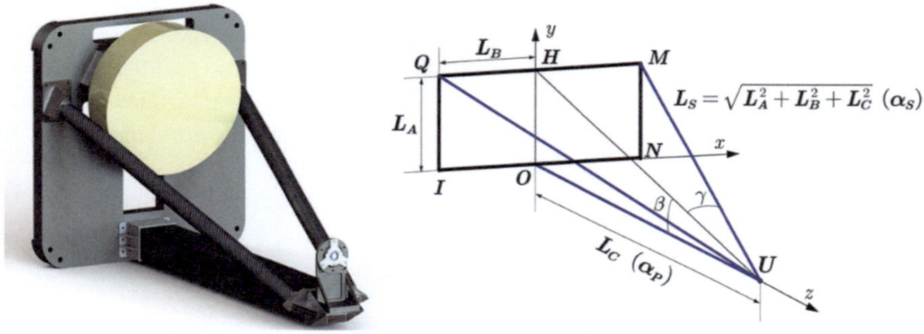

(a) (b)

Fig. 26.1 Solid model **a** Theoretical Model (**b**)

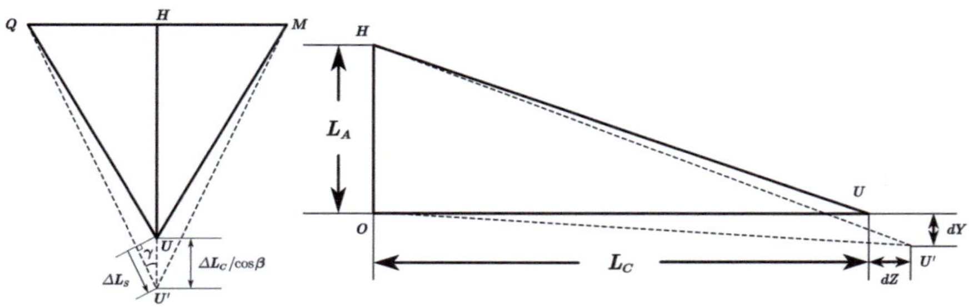

(a) (b)

Fig. 26.2 Projections of the truss model on the QMU **a** and YOZ **b** planes

$$\begin{cases} L_S = \sqrt{L_A^2 + L_B^2 + L_C^2} \\ \cos\beta = L_C / \sqrt{L_A^2 + L_C^2} \\ \cos\gamma = \sqrt{L_A^2 + L_C^2} / \sqrt{L_A^2 + L_B^2 + L_C^2} \\ \Delta L_S = \frac{\Delta L_C}{\cos\beta} \times \cos\gamma \end{cases} \tag{26.1}$$

The basic equation for thermal expansion is $\Delta L = L \times \alpha \times \Delta T$, which is inserted into the Eqs. (26.1) to obtain the thermal deformation of the node U as a function of the geometric parameters of the structure:

$$dy = \frac{L_A^2 + (L_C + \alpha_P L_C \Delta T)^2 - \left(L_S \times \cos\gamma + \frac{\alpha_S L_S \Delta T}{\cos\gamma}\right)^2}{2L_A} \tag{26.2}$$

$$dz = \sqrt{(L_C + L_C \alpha_P \Delta t)^2 - (dy)^2} - L_C \tag{26.3}$$

In the general design of truss structures, the same material is typically chosen for all support rods, denoted as $\alpha_s = \alpha_P$. However, it has been found that the structural thermal deformation, dy, can be reduced by assigning different values such that $\alpha_S \neq \alpha_P$, as demonstrated by Eq. (26.3). The principle of the optimized design is shown in Fig. 26.3.

Based on the current design of the telescope prototype, the basic structural design parameters are as follows: $L_A = 326$ mm, $L_B = 215$ mm, $L_C = 838$ mm, and the global space temperature change of the telescope during the orbital phase is $\Delta T = 1$ K.

Substituting the above structural parameters into Eq. (26.2), it is found that when $\alpha_P : \alpha_S = 1.217 : 1$, dy = 0. To meet the thermal expansion coefficient requirements of the main support rod and the side support rods, an in-depth discussion is needed on the relationship between the lay-up method of CFRP material and the CTE.

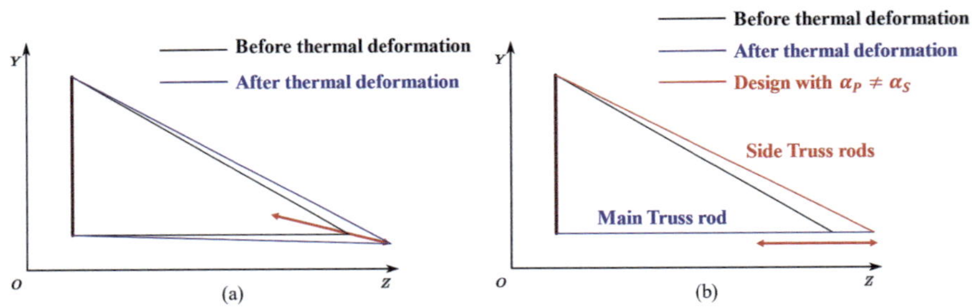

Fig. 26.3 The schematic diagram of differentiated structure design principles: $\alpha_S = \alpha_P$ (**a**) and $\alpha_S \neq \alpha_P$ (**b**)

26.2.2 Mathematical Modeling of Thermal Deformation of CFRP

CFRP is a lamination material, and its performance completely depends on the fiber direction of each layer, including thermal expansion properties. The classical composite material mechanics theory has the stress–strain relationship:

$$\begin{bmatrix} N \\ M \end{bmatrix} = \begin{bmatrix} A & B \\ B & D \end{bmatrix} \begin{bmatrix} \varepsilon^0 \\ \kappa \end{bmatrix} \tag{26.4}$$

In this context, $\boldsymbol{N}$ and $\boldsymbol{M}$ represent the stress and moment per unit length experienced by the material, while ε^0 and κ are the linear strain and curvature in the composite material layers, respectively. $\boldsymbol{A}$, $\boldsymbol{B}$ and $\boldsymbol{D}$ are the elastic stiffness matrix, coupling stiffness matrix, and bending stiffness matrix, respectively. When the layup is symmetrical and balanced, the matrix $\boldsymbol{B} = 0$. The thermal expansion does not involve bending stress, which means it does not involve calculations with the $\boldsymbol{D}$ matrix. Therefore, Eq. (26.4) can be simplified and inverted:

$$\begin{bmatrix} \varepsilon_x^0 \\ \varepsilon_y^0 \\ \gamma_{xy}^0 \end{bmatrix} = \begin{bmatrix} A'_{11} & A'_{12} & A'_{16} \\ A'_{12} & A'_{22} & A'_{26} \\ A'_{16} & A'_{26} & A'_{66} \end{bmatrix} \begin{bmatrix} N_x^T \\ N_y^T \\ N_{xy}^T \end{bmatrix} \tag{26.5}$$

If the laminates have n layers, then the relationship between the transformed force vector $\left[N^T\right]$ and each layer's stiffness matrix $\left[\overline{Q}\right]_k$, the temperature change ΔT, the thermal expansion coefficient $[\alpha]_k$ under the general coordinate system, and the thickness of the layer $(z_k - z_{k-1})$ is as follows:

$$\left[N^T\right] = \begin{bmatrix} N_x^T \\ N_y^T \\ N_{xy}^T \end{bmatrix} = \Delta T \sum_{k=1}^{n} \left[\overline{Q}\right]_k \begin{bmatrix} \alpha_x \\ \alpha_y \\ \alpha_{xy} \end{bmatrix}_k (z_k - z_{k-1})(k = 0, 1, 2, \ldots, n) \tag{26.6}$$

Meanwhile, the relationship between the elements of the tensile stiffness matrix $[A]$ and each layer's stiffness matrix, temperature change, CTE under the general coordinate system, and the thickness of the layer is as follows:

$$A_{ij} = \sum_{k=1}^{n} \left(\overline{Q}_{ij}\right)_k (z_k - z_{k-1}) \tag{26.7}$$

Substituting Eqs. (26.6) and (26.7) into Eq. (26.5), and canceling out the temperature difference ΔT, the relationship between the thermal expansion coefficient $[\alpha]_l$ of the laminated plate and the properties, ply angles, and thicknesses of each layer can be obtained as follows:

$$[\alpha]_l = [A]^{-1} \sum_{k=1}^{n} [t]_k^{-1} (z_k - z_{k-1}) \left([Q]_k [\alpha]_k \right) \quad (26.8)$$

From Eq. (26.8), it can be understood that once the single-layer plate is determined, the stiffness matrix and thickness of the material are also determined. The only variable is the single-layer plate angle transformation matrix $[t]$, indicating that the thermal expansion coefficient $[\alpha]_l$ of the laminated plate can be adjusted through the fiber ply angles of each layer in the lamination.

26.2.3 Mathematical Modeling of Thermal Deformation of CFRP

From the previous analysis, it is known that the optimal ratio of the CTE between the side support rods and the main support rod is 1:1.217. Consequently, a targeted ply design is necessary. However, finding the optimal layup design that meets the required is a key issue.

It is also important to consider that CFRP materials are anisotropic, which means that the CTE in different directions are distinct. While ensuring that the axial thermal deformation of the structure meets the requirements, it is also necessary to optimize the radial thermal expansion to match the metal joints. This is to avoid excessive thermal stress between different materials, which could negatively impact performance (Fig. 26.4).

Therefore, the optimization objectives for the CFRP layup are threefold:

(1) Ensure that the axial thermal deformation ratio of the main and side rods is approximately 1.217:1;
(2) Ensure that the axial thermal deformation coefficient is less than 0.5×10^{-6}/K, with a smaller value representing a more optimal outcome;

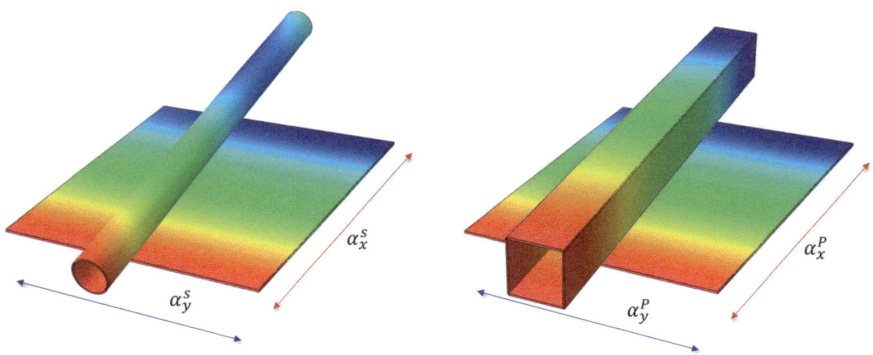

Fig. 26.4 Schematic diagram of CFRP optimization objectives: α_x^S, α_y^S, α_x^P, α_y^P with side rods (**a**) main rod (**b**)

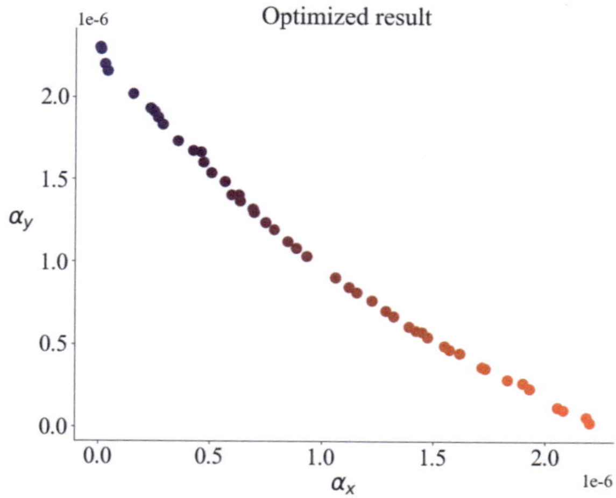

Fig. 26.5 The schematic diagram of CTE results after NSGA-II optimization

(3) Ensure that the radial thermal deformation coefficient is close to 1.0×10^{-6}/K of the invar material to avoid excessive thermal stress at the connection points.

It is therefore evident that the layup design for CFRP constitutes a multi-objective, multi-variable optimization problem. The most mature method for solving such problems is the Non-dominated Sorting Genetic Algorithm-II (NSGA-II) algorithm. The NSGA-II is a genetic algorithm based on the principle of non-dominated sorting [8, 9]. The algorithm's special non-dominated sorting and crowding distance calculation endow it with good stability and distribution, making it a fundamental algorithm often used for comparison with other optimization algorithms.

The NSGA-II algorithm, when combined with the thermal deformation theory of CFRP materials, can be utilized to optimize the fiber angles for the CFRP layup of the main support rod and side support rods. This approach will yield a multitude of layup optimization results that simultaneously fulfill the three design objectives. Subsequently, the most suitable layup scheme can be selected from the results and subjected to further analysis within the structure (Fig. 26.5).

26.3 Results

The optimized layup is presented in Table 26.2. Among them, Layup1 is a conventional quasi-isotropic layup, serving as the basic control group. The first three layups have angle constraints and are conventional layups, with good processability but larger axial thermal expansion; layups 5–8 are unconventional angle layups, with axial thermal expansion coefficients meeting design requirements. It is observed that both layup 6 and layup 7 have axial CTE less than 0.5×10^{-6}/K, and their ratio meets the 1:1.217 requirement. Layup 8 exhibits a notably low axial CTE, yet its radial CTE is relatively high, rendering it unsuitable for adoption.

Table 26.2 Different layups and CTE performance results

Layup	Ply staking sequence	α_x $(10^{-6}/\text{K})$	α_y $(10^{-6}/\text{K})$
1	$[45/-45/0/90/90/0/-45/45/90/0]_s$	1.439	1.439
2	$[45/-45/0/90/0/60/-60/0/90/0]_s$	1.103	1.843
3	$[45/-45/0/60/-60/0/90/-45/45/0]_s$	0.9777	1.993
4	$[37.08/\text{-}59.45/31.38/\text{-}57.01/16.64/\text{-}16.64/57.01/\text{-}31.38/59.45/\text{-}37.08]_s$	0.5314	2.601
5	$[32.61/\text{-}53.67/37.04/\text{-}56.86/32.65/\text{-}32.65/56.86/\text{-}37.04/53.67/\text{-}32.61]_s$	0.5242	2.502
6	$[30.49/\text{-}51.63/36.10/\text{-}60.38/30.44/\text{-}30.44/60.38/\text{-}36.10/51.63/\text{-}30.49]_s$	0.4338	2.662
7	$[40.56/\text{-}37.26/50.84/\text{-}25.04/55.81/\text{-}55.81/25.04/\text{-}50.84/37.26/\text{-}40.56]_s$	0.3548	2.736
8	$[13.40/\text{-}10.49/13.98/\text{-}65.31/1.055/\text{-}13.40/10.49/\text{-}13.98/65.31/\text{-}1.055]_s$	0.0032	6.443

Prior to considering structural thermal deformation, it is essential to ascertain that the structure is capable of functioning under conventional conditions. This entails a comprehensive assessment of the structural strength, stiffness, and overall structural weight. As demonstrated in Table 26.3, the structure satisfies the present structural requirements in four mechanical properties, indicating that the utilization of CFRP as the truss structure layup scheme can fulfill the fundamental structural strength, stiffness, and other conventional requirements.

The incorporation of layups 6 and 7 into the finite element simulation is described herein. In the simulation model, the connection point for the secondary mirror is identified as point U in Fig. 26.2. The resulting thermal deformation of the structure is as follows (Fig. 26.6):

To verify the significance of the optimization results, a comparison is made between the truss with differentiated layup designs and the one with quasi-isotropic layups for truss structure. Furthermore, the third row represents theoretical truss

Table 26.3 Summary of structural mechanical properties

Specification	Requirement	Result
1.Mass with 3 support rods	< 7.5 kg	6.37 kg
2.First eigenfrequency	> 80 Hz	141.66 Hz
3.Deformation due to gravity	$\Delta z_G < 10 \mu m$ $\Delta y_G < 10 \mu m$	$\Delta z_G = 0.31 \mu m \Delta y_G = 8.81 \mu m$
4.Strength requirements	σ_{Max} <35MPa	4.96 MPa

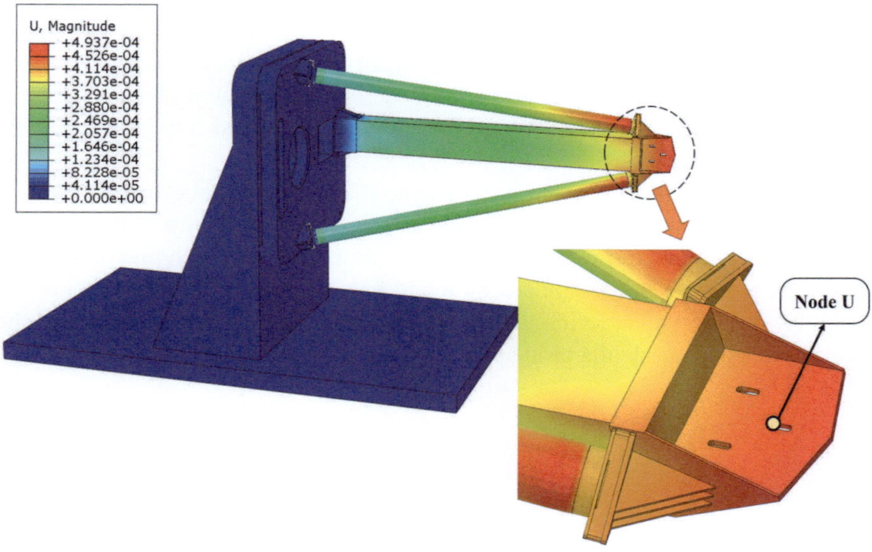

Fig. 26.6 Structural thermal deformation after bringing experimental results into finite element simulation

Table 26.4 Results of structure thermal deformation for different cases

	Structure type ($10^{-6}/K$)	dy/K (nm)	dz/K (nm)
1	Unoptimized truss structure with layup1	423.2	1198.4
2	Structure simulation result with layup optimization	4.229	454.7
3	Truss theoretical calculation results	3.416	363.5

values derived from Eqs. (26.2) and (26.3), which are used to validate the simulation results (Table 26.4):

The results of the finite element simulation closely correspond with the theoretical calculations for the truss. Furthermore, it has been observed that, compared to the conventional scheme without layup optimization, the optimized scheme with differentiated layup designs for the side support rods and main support rod significantly reduces the thermal deformation of dy while ensuring that dz remains unchanged.

26.4 Conclusion and Discussion

This study proposes an optimized design method that combines truss structure with the characteristics of CFRP layup, targeting the thermal deformation requirements in the dy and dz directions for gravitational wave detection telescopes. Overall, based on truss theory modeling, the optimal CTE ratio of the main support rod to the side support rods is obtained, followed by the use of the NSGA-II algorithm to achieve a CFRP layup that meets multiple objectives. The optimized layup is then incorporated into the structure and compared with theoretical calculations and non-optimized schemes. The results show that the solid element finite element simulation is close to the theoretical calculation results, and the thermal deformation of the structure is significantly reduced compared to the conventional design, indicating a remarkable optimization effect.

It should also be noted that the current layup selection does not constrain the fiber angles of each layer, belonging to an unconventional angle layup scheme, which poses greater manufacturing challenges and requires high precision in error control. Therefore, future research should continue to focus on the precision control of the CFRP layup process, exploring the effects of layup errors on the structure and the requirements of the manufacturing process.

References

1. W. Fan et al., The impact of telescope aberrations on the magnitude of tilt to length coupling noise in space based gravitational wave detectors. Class. Quantum Gravity **40**, 245007–245007 (2023)
2. V.K. Milyukov, TianQin space-based gravitational wave detector: key technologies and current state of implementation. Astron. Rep. **64**(12), 1067–1077 (2020)
3. Y. Yan et al., Optimized design of a gravitational wave telescope system based on pupil aberration. Appl. Opt. **63**, 1815–1815 (2024)
4. F. Kazemi, T. Shafighfard, D.-Y. Yoo, Data-driven modeling of mechanical properties of fiber-reinforced concrete: a critical review. Arch. Comput. Methods Eng. **31**(4), 2049–2078 (2024)
5. J.C. Livas et al., Telescopes for space-based gravitational wave missions. Opt. Eng. **52**, 091811–091811 (2013)
6. A. Uminska et al., Ground testing of the LISA telescope. Astron. Opt.: Des, Manuf, Test Space Ground Syst. **III**, 11820 (2021)
7. J. Sanjuán et al., Carbon fiber reinforced polymer dimensional stability investigations for use on the laser interferometer space antenna mission telescope. Rev. Sci. Instrum. **82**, 124501 (2011)
8. K. Deb, A. Pratap, S. Agarwal, T. Meyarivan, A fast and elitist multiobjective genetic algorithm: NSGA-II. IEEE Trans. Evol. Comput. **6**, 182–197 (2002)
9. L. Aydin, H.S. Artem, H.A. Deveci, Single and multiobjective optimizations of dimensionally stable composites using genetic algorithms. Mech. Compos. Mater. **57**(3), 321–336 (2021)

Chapter 27
Design of a μK-Level High-Precision Temperature Sensing System for TianQin Inertial Sensor Thermal Diagnostics

Hui Chen, Xing-Yu Yan, Shao-Bo Qu, Yanzheng Bai, and Ze-Bing Zhou

Abstract The space gravitational wave detection TianQin mission requires that the residual acceleration noise of the inertial sensor does not exceed 10^{-15} m/s^2/Hz$^{1/2}$ at around 6 mHz in the sensitive axis. The temperature gradient noise is one of the main direct disturbance effects on the test mass. The TianQin satellite demands that the temperature difference noise on both sides of the sensitive axis of the inertial sensor test mass be controlled within 5 μK/Hz$^{1/2}$ in the measurement frequency band. This work introduces the design scheme of a μK-level high-precision temperature sensor, covering a temperature measurement range of 10 ~ 30 °C, with a temperature resolution of 3 μK/Hz$^{1/2}$ at a frequency of 6 mHz. A temperature sensing circuit based on AC carrier modulation of the Wheatstone bridge is proposed. Due to the potential influence of ground environment temperature on the performance testing of high-precision temperature sensors, an ultra-high stability thermal insulation platform is designed for the noise test and evaluation on the μK-level.

27.1 Introduction

Temperature is one of the seven basic physical quantities in the International System of Units, and there is a high-precision temperature measurement requirement in many precise measurement experiments. Among them, in space gravitational wave detection, the temperature sensing technology used by LISA [1], which employs thermal resistance and AC bridge technology at room temperature, has reached a testing level of 20μK/Hz$^{1/2}$ at 0.1mHz [2]. In the gravity gradient mission GOCE, a platinum resistance thermometer and a thermocouple bridge circuit is used, which achieved a temperature accuracy of 5μK/Hz$^{1/2}$ at 5mHz [3]. Integrated optical temperature

H. Chen · X.-Y. Yan · S.-B. Qu (✉) · Y. Bai · Z.-B. Zhou
National Gravitation Laboratory, MOE Key Laboratory of Fundamental Physical Quantities Measurement, and School of Physics, Huazhong University of Science and Technology, Wuhan, People's Republic of China
e-mail: qushaobo@hust.edu.cn

J. Luo et al. (eds.), *Proceedings of the 7th International Workshop on the TianQin Science Mission; 25–26 April 2024, Hong Kong SAR, China*, Springer Proceedings in Physics 428, https://doi.org/10.1007/978-981-96-7803-7_27

sensors nowadays have attracted extensive interests thanks to its merits of high accuracy, which could be better than 60mK in a measurement range of about 297 ~ 415K [4].

Thermal noise and the thermal gradient noise [5] are important noise sources of inertial sensor and the laser interferometry system for space gravitational detection [6].

The requirement for the acceleration noise spectral density in the sensitive axis direction of the inertial sensor in the TianQin mission is below 1×10^{-15} m/s^2/Hz^1/^{2}, which is more stringent than that of the LISA inertial sensor. The TianQin satellite [7] demands that the temperature difference noise on both sides of the sensitive axis of the inertial sensor test mass be controlled within 5μK/Hz$^{1/2}$ within the measurement frequency band. The thermal effects are also the main error source for many other key payloads and components, like the frequency stabilized laser [8], the telescope structure, etc. These payloads also need a stable thermal environment at μK level.

Currently, high-precision temperature control remains a key technical challenge of TianQin, therefore designing a high-precision temperature sensor capable of detecting temperature changes at μK level is essential. According to the requirement of TianQin mission to detect temperature fluctuations at the critical points of the key payloads, the temperature sensor should be able to have a measurement range of 10 ~ 30 °C, with a temperature resolution of 3 μK/Hz$^{1/2}$ at a frequency of 6 mHz.

27.2 Schematic Design

To meet the overall needs of resolution, this article proposes a temperature sensing solution based on a Wheatston bridge-type resistive [9] structure modulated by high precision AC carrier. A squarewave carrier modulation is used to modulate the low-frequency temperature signals to higher frequencies, thereby enhancing the signal-to-noise ratio of low-frequency signals and improving the resolution. The schematic design is shown in Fig. 27.1.

The circuit output expression is:

$$V_o = V_{ref}\left(\frac{R_2}{R_1 + R_2} - \frac{R_{TS}(T)}{R_3 + R_{TS}(T)}\right) \times G \tag{27.1}$$

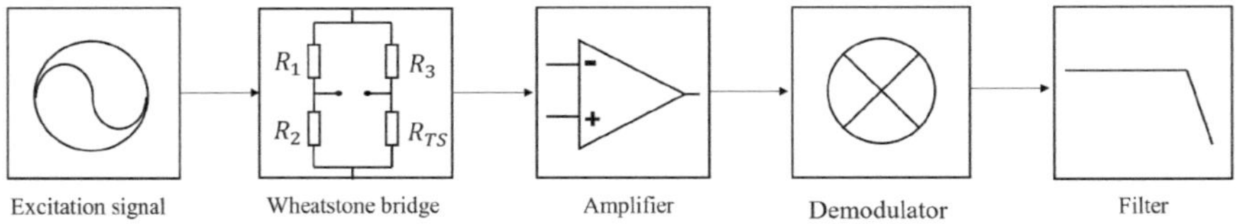

Fig. 27.1 Schematic diagram of the overall principle of temperature sensor circuit

where V_{ref} is the electrical bridge excitation signal, R_{TS} is thermistor and G represents the gain of the circuit's output stage amplifier.

The main principle of the circuit is that the temperature sensor resistance converts temperature signals into electrical signals, the AC carrier modulates the low-frequency temperature signals to the bridge, and then the signals are amplified through an instrumentation amplifier. The temperature signal is extracted using a demodulator, and the temperature variation characteristics are obtained through the acquisition system.

As the excitation signal, the noise of the squarewave is one of the main noise sources. Therefore, the carrier source must be specially designed to eliminate the overall noise level of the circuit. Based on this, a high-stability voltage reference scheme is proposed by using multi reference sources in parallell. Theoretical analysis shows that by paralleling multiple high-precision voltage reference components, with number N, the carrier wave amplitude stability can be improved by a factor of about $N^{1/2}$. In order to meet the stability requirement, a number of 10 is used in this paper.

To reduce the circuit background noise, a phase sensitive demodulation scheme is designed. The traditional scheme of the temperature sensor developed by the LISA team is also based on the Wheatston bridge and AC modulation scheme, but its demodulation circuit adopts the method of high-speed sampling and digital demodulation [10, 11], which may increase the complexity and power consumption. To meet the demands of low power consumption and ability of integrated design for space applications, the design scheme proposed in this article adopts a circuit scheme based on analog switch demodulation, using as few electronic components as possible to achieve temperature measurement requirement.

Due to the excellent low-frequency performance of the circuit structure, the background noise of the circuit is expected to be reduced to meet the stringent requirement of TianQin mission.

27.3 Noise Analysis

To assess the noise model and the contribution of each noise sources, we analyzed the equivalent temperature noise from several main sources. Its expression is as follows:

$$T_{\text{n\_ system}} = \frac{\delta V_o(\text{V}/\sqrt{\text{Hz}})}{\partial V_o / \partial T(\text{V/K})} = [T^2_{\text{n\_}\,V\text{ref}} + T^2_{\text{n\_R}} + T^2_{\text{n\_}G} + T^2_{\text{n\_el}}]^{1/2} \tag{27.2}$$

where $T_{\text{n\_}V\text{ref}}$ is the noise terms introduced by the fluctuations in carrier amplitude, $T_{\text{n\_R}}$ is the noise introduced by resistance value fluctuations, $T_{\text{n\_}G}$ is the amplifer gain fluctuations introduced noise and $T_{\text{n\_el}}$ is the inherent noise of the sensing circuit.

According to the existing design, the noise model of the system is analysed. Based on the specific parameters, the noise equivalent temperature (NET) curves for each items are calculated, as shown in the Fig. 27.2.

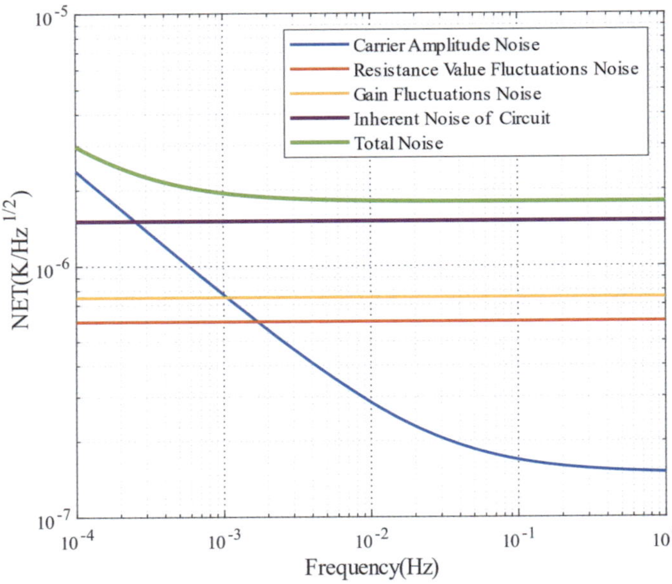

Fig. 27.2 Noise equivalent temperature power spectrum curves

As the combined noise curve shows in the Fig. 27.2, the total noise spectrum density is lower than 2 $\mu K/Hz^{1/2}$ at 6 mHz, which indicates that this design theoretically meets the requirements for TianQin inertial sensor temperature measurement and controlling.

27.4 Thermal Insulation Testing Platform Design

To evaluate the noise of the temperature measurement system, the actual temperature fluctuation in the environment where the temperature sensor is under test must be significantly smaller than 3 $\mu K/Hz^{1/2}$. Therefore, the requirement we put forward for the test platform is as follows:

$$T_{\text{testbed}} \leq 10^{-6}(\text{K}/\sqrt{\text{Hz}}) \tag{27.3}$$

In order to meet this requirement, we preliminarily design the test platform to be a three-layer thermal insulation sphere, and the specific illustration is shown in Fig. 27.3.

For the structure in Fig. 27.3, the temperature distribution satisfies the Fourier partial differential equation [12]:

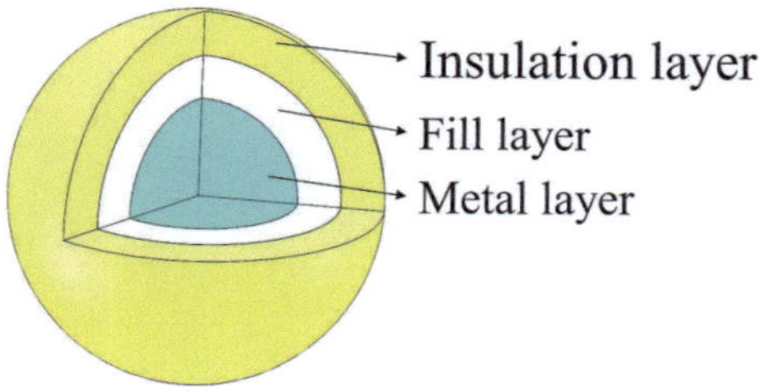

Fig. 27.3 Multi-layer thermal insulating platform diagram

$$\rho c \frac{\partial T(r,t)}{\partial t} = k \nabla^2 T(r,t) \tag{27.4}$$

where ρ, c, and k are the density, the specific heat and the thermal conductivity, respectively.

In order to obtain the frequency domain thermal transfer function, a Laplace transform is applied to the equation, resulting in the following expression:

$$sU_i(r,s) - u_i(r,0) = \frac{q_i^2}{s} \frac{d^2 U_i(r,s)}{dr^2} \tag{27.5}$$

where $U(r,s) \equiv \mathcal{L}[u(r,t)]$, $U_i(r,0) = 0$ (when $i = 1$, it corresponds to the innermost structure). Considering the continuity of temperature and heat flux density at the boundary, the solution can be obtained.

The transfer function expression for the temperature transfer from the outside to the center of the test bench is:

$$\lim_{r \to 0} H_{\text{ins}}(r,s) = \lim_{r \to 0} [U_1(r,s)/r] \Big/ T_{\text{amb}}(s) \tag{27.6}$$

where $T_{\text{amb}}(s) \approx 0.1\text{K}/\sqrt{\text{Hz}}$ is the environmental temperature fluctuation. Therefore, it is easy to set the requirements for the transfer function of the insulator platform:

$$|H_{\text{ins}}(\omega)| \leq 10^{-5} \tag{27.7}$$

Corresponding to Fig. 27.3, once the specific parameters that satisfies Eq. (27.7) are introduced, we can obtain the theoretical transfer function curve in Fig. 27.4, which meets the requirements and the platform can be used for testing and verifying the performance of the temperature sensor.

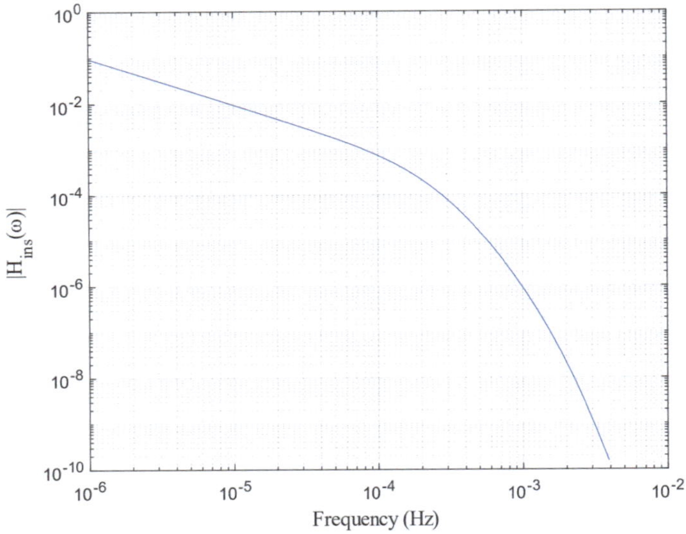

Fig. 27.4 The theoretical transfer function curve

27.5 Discussion

In order to satisfy the requirements of the TianQin inertial sensor thermal diagnostics, a high-precision temperature sensor is designed. To meet the high precision and low noise requirements of the temperature sensor, an AC squarewave carrier modulation scheme is chosen to suppress the low-frequency noise of the amplifier. After completing the overall design, the theoretical noise curve is analysed and presented, indicating that the design requirements are met. To verify the performance of the temperature sensor, a thermal insulated platform is proposed and designed, the thermal isolation ratio meets the testing requirements of the temperature sensor in this paper. The next step will be mainly on the performance testing and validation of the theoretical design.

Acknowledement This work was supported by the National Key R&D Program of China (No. 2022YFC2204404).

References

1. P. Amaro-Seoane, H. Audley, S. Babak, J. Baker, E. Barausse, P. Bender, E. Berti, P. Binetruy, M. Born, D. Bortoluzzi, J. Camp, Laser interferometer space antenna (2017).arXiv:1702.00786
2. D. Roma-Dollase, V. Gualani, M. Gohlke et al., Resistive-based micro-kelvin temperature resolution for ultra-stable space experiments. Sensors **23**(1), 145 (2023)

3. D. Valentini, M. Vacance, D. Battaglia, GOCE instrument thermal control. SAE Trans **115**, 115–126 (2006)
4. Y.M. Wang, H.W. Shu, X.Y. Han, Research on high-precision silicon-based integrated optical temperature sensor. Proc. SPIE **12062**, 120620L (2021)
5. G.L. Li, J. Ke, H.G. Li et al., Thermal induced noise on test mass with copper alloy electrode housing for spaceborne gravitational wave detection. Phys. Rev. D **109**(8), 082001 (2024)
6. J.J. Mao, Y.J. Tan, J.P. Liu, S. Yang, J. Luo, C.G. Shao, Z. Zhou, Residual gas damping noise in constrained volume in space-borne gravitational wave detection. Class. Quant. Grav **40**(7), 075015 (2023)
7. J. Luo, Y.Z. Bai, L. Cai et al., The first round result from the TianQin-1 satellite. Class. Quant. Grav **37**(18), 185013 (2020)
8. Y. Luo, H. Li, Y.Q. Li, Z.H. Zhan, H.C. Yeh, Prototype of a monolithic cavity-based ultrastable optical reference for space applications. Appl. Opt. **60**(10), 2877–2885 (2021)
9. P.R.N. Childs, J.R. Greenwood, C.A. Long, Review of temperature measurement. Rev. Sci. Instrum. **71**(8), 2959–2978 (2000)
10. J. Sanjuán, A. Lobo, M. Nofrarias, J. Ramos-Castro, P.J. Riu, Thermal diagnostics front-end electronics for LISA pathfinder. Rev. Sci. Instrum. **78**(10), 104904 (2007)
11. M. Armano, H. Audley, J. Baird, D. Bortoluzzi, E. Castelli, A. Cavalleri, Cesarini A, Cruise AM, K. Danzmann,Temperature stability in the sub-milliHertz band with LISA Pathfinder. Mon. Not. Roy. Astron. Soc. **486**(3), 3368–3379 (2019)
12. Ezekoye, O.A, Conduction of heat in solids. In: M.J. Hurley, et al, *SFPE Handbook of Fire Protection Engineering*. Springer, New York, NY (2016)

Chapter 28
Analysis of the Influence of Telescope Pointing Deviation Angle on the Coupling Efficiency of MMF in TianQin Laser Ranging System

Chunyu Yuan and Ming Li

Abstract The TianQin Station has applied a multimode fiber (MMF) coupled superconducting single photon detector (SNSPD) in the space target laser ranging system. The coupling manner of the space light and MMF is used in the receiving system and it can affect the number of echo photons and ranging accuracy, especially in lunar laser ranging, high orbit satellite laser ranging and space debris tracking. When there is a pointing deviation angle from the telescope to the measurement target, it will directly affect the coupling efficiency and further affect the entire system efficiency. The paper analyzes the influence of the entire optical transmission process from the 1.2m telescope receiving system to the MMF on the optical fiber coupling efficiency. It mainly includes detailed calculations of the aberration, the spot properties and reflection. The variation of MMF coupling efficiency corresponding to different telescope deviation angles for 200um multimode fiber is given.

28.1 Introduction

Space target laser ranging [1] refers to the process of emitting pulsed laser from the ground station, which is directed towards the measured target through the optical system of telescope. The weak laser signal reflected from the observed target is then incident onto the photosensitive surface of the photodetector through a receiving optical system of telescope. The TianQin Station [2] uses a superconducting single photon detector (SNSPD) with the multimode fiber transmission. The laser ranging system couples the echo spatial beam onto the detector chip using the multimode fiber (MMF). In the past few years, space optical communication and laser ranging

C. Yuan (✉) · M. Li
MOE Key Laboratory of TianQin Mission, TianQin Research Center for Gravitational Physics and School of Physics and Astronomy, Frontiers Science Center for TianQin, Gravitational Wave Research Center of CNSA, Sun Yat-Sen University (Zhuhai Campus), Zhuhai, China
e-mail: yuanchy26@mail2.sysu.edu.cn

J. Luo et al. (eds.), *Proceedings of the 7th International Workshop on the TianQin Science Mission; 25–26 April 2024, Hong Kong SAR, China*, Springer Proceedings in Physics 428, https://doi.org/10.1007/978-981-96-7803-7_28

technology based on the optical fiber have received widespread attention [3, 4]. Previous studies have mainly focused on the calculation of single-mode coupling efficiency and MMF [5, 6], but there are relatively few actual calculations for the application of MMF in the system. In fact, the ranging echo signal of space targets is too weak, and is usually in a single photon state. For the telescope with 1.2m main mirror at the TianQin Station, its pointing deviation angle has a significant impact on the fiber coupling. When there is a deviation in the angle between the telescope and the measured target, it can affect the strength of the echo signal, especially for high orbit satellites ranging, lunar laser ranging, and space debris tracking. Therefore, analyzing the deviation angle of the telescope is crucial for the coupling efficiency between space light and multimode fiber. The following analysis mainly focuses on the aberration, the light spot, and reflection losses, and finally presents the changes in MMF coupling efficiency corresponding to different telescope deviation angles.

28.2 Theoretical Calculation

28.2.1 Aberration Analysis

The aberration of the rotationally symmetric optical system can be expressed in the expansion formula of a third-order aberration function as [7]:

$$\begin{aligned} \mathrm{W}(\mathrm{H}, \rho, \phi) = W_{040}\boldsymbol{\rho}^4 + W_{131}H\boldsymbol{\rho}^3\cos\phi + W_{222}\mathbf{H}^2\boldsymbol{\rho}^2\cos^2\phi \\ + W_{220}\mathbf{H}^2\boldsymbol{\rho}^2 + W_{311}\mathbf{H}^3\rho\cos\phi + W_{020}\boldsymbol{\rho}^2 + W_{111}H\rho cos\phi + W_{200}\rho \end{aligned} \tag{28.1}$$

H is the vector of viewing field, $\boldsymbol{\rho}$ is the vector of pupil, ϕ is the angle between two vectors. When the field angle is determined, Eq. (28.7) can be transformed into the form of Seidel primary aberration function, as follows:

$$W(\rho, \phi) = a_{40}\rho^4 + a_{31}\rho^3 cos\phi + a_{22}\rho^2 cos^2\phi + a_{20}\rho^2 + a_{11}\rho cos\phi \tag{28.2}$$

When the pointing of telescope has a deviation angle, the introduced aberrations include tilt, astigmatism, and coma. The calculation of aberrations is based on the optical system transmitted from the ranging telescope to MMF. The optical parameters are listed in the Table 28.2. The calculation results are in Fig. 28.1. As shown in the figure, when the deviation angle reaches 12 arcsec, the RMS of maximum tilt is 9.7 λ, the RMS of maximum astigmatism is 0.1582 λ, and the RMS of maximum coma is 0.142 λ.λ.

Table 28.2 Table of the optical parameters

Surface	Type	Conic constant	Focal length
Main mirror	Conic	–1.018	1800
Secondary mirror	Conic	–2.058	–538.043
Reflector	Flat		
Coupling lens	Double bonded lens		37.7

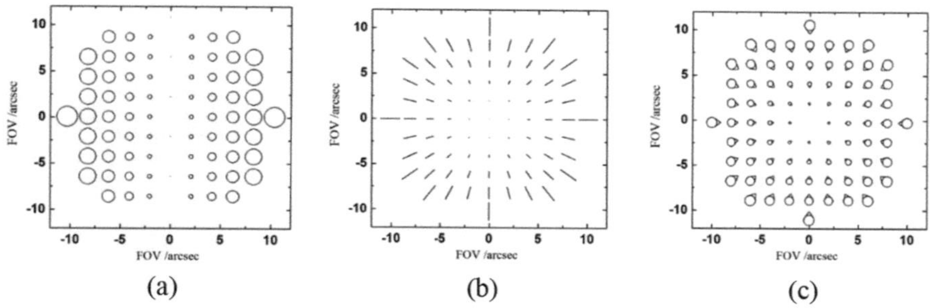

Fig. 28.1 The change of aberration: **a** tilt, **b** astigmatism, **c** coma

28.2.2 Spot Analysis

When there is a telescope pointing deviation angle from optical axis, the coefficient of off-axis aberration is generated by optical system transmission. The properties of the light spot transmitted to the front surface of the optical fiber change, including the the light field intensity, phase, and position. It is considered that the beam entering the telescope is parallel light, so the laser wavefront intensity of the incident light can be expressed as [8]:

$$E_A(r) = E_S exp\big[-j\phi(r,\theta)\big] \tag{28.3}$$

E_S is the wavefront intensity before generating the deviation angle. When the beam is transmitted to the coupling lens, the phase change is expressed as [8]:

$$E_1(r,\theta) = E_S exp\big[-j\phi(r,\theta)\big] exp(-\tfrac{\pi}{\lambda f}(r^2)] \tag{28.4}$$

$\exp(-\frac{\pi}{\lambda f}(r^2)]$ is the phase factor of the coupling lens.

The beam is focused onto the receiving field of MMF through the coupling lens, which conforms the theory of angular spectrum transmission. If the distance from the back surface of the coupling lens to the optical fiber front surface is d, the energy distribution at the optical fiber front surface can be expressed as:

$$E_A(x,y) = \mathcal{F}^{-1}\big\{\mathcal{F}\{E_1(x,y)\}H\big(f_x, f_y, d\big)\big\} \tag{28.5}$$

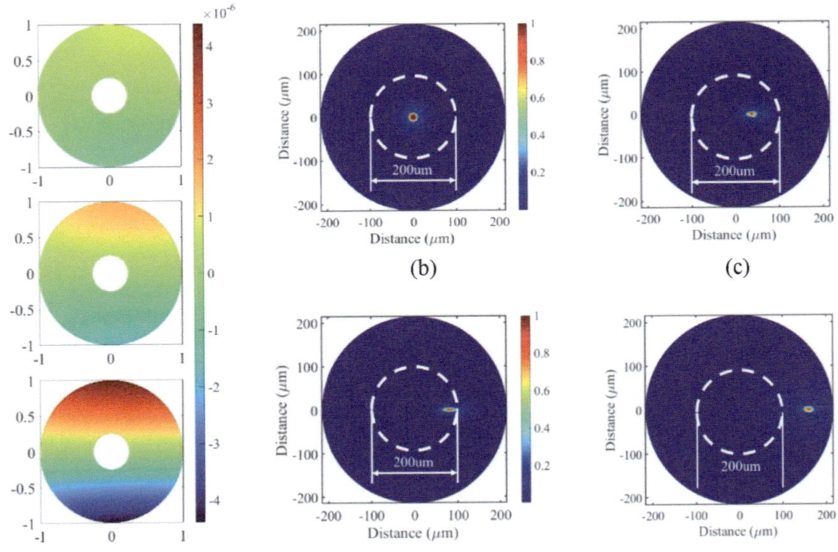

Fig. 28.2 The change of optical field intensity and phase. **a** phase: 2 ", 4", and 8"(from top to bottom). Optical field intensity: **b** 0 ", **c** 2", **d** 4", **e** 8"

$H(f_x, f_y, d)$ is transfer function, and it can be expressed as:

$$H(f_x, f_y, d) = F\{h(x, y, d)\} = exp\left(ikd\sqrt{1 - (\lambda f_x)^2 - (\lambda f_y)^2}\right) \tag{28.6}$$

f_x and f_y are the spatial frequencies.

Based on the above analysis, the optical field intensity and phase of the beam transmitted to the fiber front surface can be obtained from formula (28.3) and (28.4), as shown in Fig. 28.2.

As shown in the figure, the aberration generated by the optical path not only changes the phase and optical field intensity, but also changes the position of the beam focusing on the fiber front surface. Figure 28.2 reflects the position relationship between the focusing light spot and the fiber front surface. When the deviation angle of telescope increases from 0" to 4 ", the position of the light spot on the fiber front surface is shifted to the edge of the fiber core diameter of the receiving front face. When the deviation angle of telescope increases from 4 "to 6", the light spot intersects with the edge of the MMF fiber core, and the intensity changes dramatically. Until it reaches 8", the spot is completely removed from the fiber core.

28.2.3 Reflection Loss Analysis

The deviation angle of telescope causes the change in the edge angle of the focusing beam at the fiber front surface. In fact, based on the analysis of the core diameter and numerical aperture of MMF, the edge angle still does not exceed the maximum receiving critical angle of MMF before the light spot completely moves out of the fiber front surface. However, when the beam is coupled with the fiber front surface, the interface causes Fresnel reflection loss due to different refractive indices, which affects the energy of the beam coupled into the fiber core. The formula for the reflectivity of the surface is as follows:

$$a_s = \frac{4}{\pi R^2}\int_0^i \left(\frac{n_1 cos\varphi - \sqrt{n_1^2 - n_1^2 sin^2\varphi}}{n_1 cos\varphi - \sqrt{n_2^2 - n_1^2 sin^2\varphi}}\right)^2 \sqrt{\omega^2 - f^2 tg^2\varphi}\frac{f}{cos\varphi}d\varphi \tag{28.7}$$

$$a_p = \frac{4}{\pi R^2}\int_0^i \left(\frac{n_2^2 cos\varphi - \sqrt{n_1^2 - n_1^2 sin^2\varphi}}{n_2^2 cos\varphi + \sqrt{n_2^2 - n_1^2 sin^2\varphi}}\right)^2 \sqrt{\omega^2 - f^2 tg^2\varphi}\frac{f}{cos\varphi}d\varphi \tag{28.8}$$

Among them, R is the radius of the laser beam on the coupled lens surface, n_1 is the air refractive index, n_2 is the refractive index of the optic fiber core, φ the angle incidence of the edge light which is transmitted to the fiber front surface. According to the formula, the edge light affects the reflection loss rate of the beam reaching the surface. Therefore, for the coupling efficiency of MMF in the ranging telescope, the change of light spot displacement and the edge light are considered.

28.2.4 Coupling Efficiency Analysis

The coupling efficiency of MMF is the ratio of the power entering the fiber core to the power reaching the plane of the fiber front surface after passing through the coupling lens, it can be expressed as [9]:

$$\eta = \frac{\langle P_{c_i}\rangle}{\langle P_{r_i}\rangle} = \frac{\langle |E_A(r)F_A^*(p)dp|^2\rangle}{\left\langle \int_R |E_{nm}(p)|^2 dp\right\rangle} \tag{28.9}$$

P_{c_i} is the optical power received actually by the optical fiber core, P_{r_i} is the optical power of the beam to reach the fiber front surface, all of the power enter the optical fiber core. $F_A(\mathbf{p})$ is the normalized optical fiber mode, and it can be expressed as:

$$F_A(\mathbf{p}) = \sqrt{\frac{2}{\pi\omega_a^2}}exp\left(-\frac{\mathbf{p}^2}{\omega_a^2}\right) \tag{28.10}$$

w_a is mode-field radius of MMF, $p = (P_X, p_Y)$ is the transverse receiver plane coordinate.

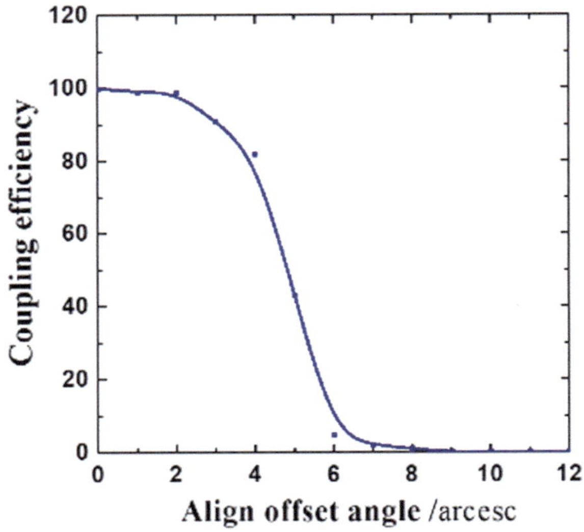

Fig. 28.3 The variation of MMF coupling efficiency with the deviation angle

Therefore, for the coupling efficiency of MMF in the ranging telescope, the change of light spot displacement and the edge light are considered. Finally, based on formula (28.9) and (28.10), the variation of MMF coupling efficiency with the deviation angle is calculated, as shown in Fig. 28.3. When the deviation angle of telescope increases from 0" to 4 ", the coupling efficiency decreases significantly. When the deviation angle of telescope increases from 4 "to 6", the coupling efficiency decreases at an accelerated rate. Until it reaches 8", the coupling efficiency approaches 0.

28.3 Results

The 1.2m main mirror telescope and MMF coupled SNSPD are used in the ranging system of TianQin Station. It is beneficial for the performance of the ranging system, but the pointing deviation angle of the telescope has the influence on the fiber coupling efficiency. The fiber coupling efficiency can affect the number of echo photons and ranging accuracy, especially in lunar laser ranging, high orbit satellite laser ranging and space debris tracking. Therefore, the paper analyzes the influence of the entire optical transmission process from the 1.2m telescope receiving system to the MMF on the optical fiber coupling efficiency. It mainly includes detailed calculations of the aberration, the spot properties and reflection. The variation of MMF coupling efficiency corresponding to different telescope deviation angles for 200um multimode fiber is given. When the deviation angle of telescope increases from 0" to 4 ", the coupling efficiency decreases significantly. When the deviation angle of telescope increases from 4 "to 6", the coupling efficiency decreases at an accelerated rate. Until it reaches 8", the coupling efficiency approaches 0.

References

1. c. Courde, J.M. Torre, E. Samain, Lunar laser ranging in infrared at the Grasse laser station. Astron. Astrophys. **602**, 12 (2017)
2. T. Gao, C. Zhang, M. Li et al., Research and experiment on lunar laser ranging at Sun Yat sen university. Acta Sci. Natur. Univ. Sunyatseni **60**(Z1), 247–252 (2021)
3. Z. Tang, R. Zhang, S.K. Mondal,et al, Optimization of fiber-optic coupling and alignment tolerance for coupling between a laser diode and a wedged single-mode fiber. Opt. Commun. **199**(1–4), 95–101(2001)
4. J. Ma, G. Lu, S. Yu, et al, Fiber coupling efficiency for satellite and space to ground downlinks with pointing errors. Optik **202**, 163558 (2020)
5. H. Cho, W. Kim, H. Park, A simple method for calculating coupling efficiency in optical interconnection system involving various misalignment parameters. Opt. Lasers Eng. **42**(5), 503–509 (2004)
6. S. Shang, J. Zhang, Y. Qi, B. Zeng, P. Jiang, H. Yang, Fiber coupling efficiency of a Bessel-Gaussian beam received by a Cassegrain antenna under atmospheric turbulence. Appl. Opt. **61**(23), 6871–6878 (2022)
7. K.P. Thompson, T. Schmid, J.P. Rolland, The misalignment induced aberrations of TMA telescopes. Opt. Express **16**(25), 20345–20353 (2008)
8. J. Li, Z. Peng, Y. Fu, Diffraction transfer function and its calculation of classic diffraction formula. Optics communications **280**(2), 243–248 (2007)
9. Y. Chen, L. Tan, L. Zhao, et al, The influence of alignment tolerance on coupling efficiency of FSOC system based on few-mode fiber. J. Opt. (1), 22 (2020)

Chapter 29
Efficiency Assessment of Echo Reception in Laser Ranging System

Lixiang Zhou, Xida Han, and Ming Li

Abstract The efficiency of a laser ranging system directly influences its ranging capability, necessitating a reliable method for testing system efficiency. We present a testing method that assesses the efficiency of a laser ranging system using only standard stars, without the need for additional equipment. The approach leverages the capabilities of single photon detectors and measuring photon flux from stars at varying zenith angles, the impact of atmospheric extinction on system efficiency is considered. Through this method, the reception efficiency of the laser ranging system at TianQin station was tested and compared with the measured values of mirror reflectance, ensuring the requirements for performance monitoring of the laser ranging system.

29.1 Introduction

The laser ranging optical system mainly consists of two major parts: the laser emission system and the echo receiving system. The laser emission system is responsible for emitting laser pulses at a specific frequency, while the echo receiving system [1] is responsible for collecting signal photons returning to the station from the cube corner reflector (CCR), and generating signals to the high precision event timer and information processing computer. The efficiency of the system significantly impacts on the laser ranging performance of the station. The system's efficiency determines how much of the photons that reaches the telescope's entrance pupil actually makes it to the output signal. Factors such as the accumulation of dust on optical surfaces [2] and the aging of optical fibers [3] can compromise the reception efficiency of the optical system, especially in the case of geographical locations like the TianQin Station. According to the laser ranging equation, the efficiency of the echo receiving

L. Zhou (✉) · X. Han · M. Li
MOE Key Laboratory of TianQin Mission, TianQin Research Center for Gravitational Physics, School of Physics and Astronomy, Frontiers Science Center for TianQin Gravitational Wave Research Center, CNSA, Sun Yat-Sen University (ZhuhaiCampus), Zhuhai, China
e-mail: zhoulx23@mail2.sysu.edu.cn

J. Luo et al. (eds.), *Proceedings of the 7th International Workshop on the TianQin Science Mission; 25–26 April 2024, Hong Kong SAR, China*, Springer Proceedings in Physics 428, https://doi.org/10.1007/978-981-96-7803-7_29

system is proportional to the capability of the laser ranging [4, 5]. Therefore, it is essential to utilize techniques to measure the efficiency of the echo receiving system and analyze the laser ranging system's capacity for receiving echoes.

29.2 Method

When the telescope is aligned with a star, the efficiency can be deduced by comparing the measured photon flux, denoted by the actual photon flux *(n)*, and the stellar brightness (*F*). Nevertheless, due to the influence of atmospheric extinction, we cannot derive the system's efficiency exclusively from the flux of star library data and the photons observed by telescope. In this section, we delineate the process of efficiency measurement, data error analyzes, and outline fundamental principles for the measurement method.

29.2.1 Fundamental Principles

Assuming a star in a standard star library radiates an intensity $F(\mathrm{Wm}^{-2}\mu\mathrm{m}^{-1})$ in the operating wavelength range of the system, without considering atmospheric extinction and with a perfect system optical efficiency of 100%, the theoretical number of photons received per second by a telescope optical system from the star's emitted light can be calculated as follows:

$$N = \frac{F\Delta\lambda A\lambda}{hc}, \tag{29.1}$$

Here, $\Delta\lambda$ (μm) represents the bandwidth of the filter, A (m^2) represents the collecting area of the telescope, λ represents the, h and c represent the Planck constant and the speed of light respectively.

Consider the echo receiving system's optical efficiency (η) the photon number (n') received by the system can be written as follows:

$$n' = \eta N, \tag{29.2}$$

During ground-based measurements, the radiated photons from the target experience attenuation by the atmosphere before reaching the detector, as shown in Fig. 29.1. This causes the atmospheric extinction to vary with different zenith angles, with more significant atmospheric extinction at greater zenith angles. As a result, the number of photons received by ground stations from stars is affected by atmospheric extinction, which varies with different zenith angles.

Airmass (*Am*) is commonly used to describe the ratio of the atmospheric optical thickness along a specific zenith angle (z) direction to that in the zenith (0°) direction.

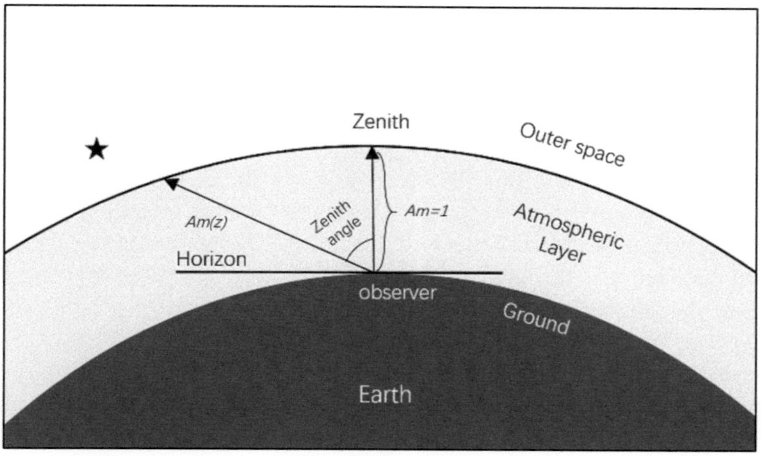

Fig. 29.1 Schematic diagram of atmospheric extinction affected by different elevation angles

It is typically calculated using empirical formulas [6] ADDIN, such as:

$$A_m = \frac{1.002432\cos^2 z + 0.148386\cos z + 0.0096467}{\cos^3 z + 0.149864\cos^2 z + 0.0102963\cos z + 0.000303978}, \quad (29.3)$$

The atmospheric transmittance (T) can be defined as:

$$T = \exp(-aA_m(z)L_0), \quad (29.4)$$

where a represents the total attenuation per kilometer.

Assuming n presents $n' - n_{noi}$. So, the photon flux from a star that can be received can be described as follows:

$$n = \eta N \exp[-\alpha L_0 A_m(z)], \quad (29.5)$$

Rearranging equation and taking the natural logarithm of both sides:

$$\ln\left(\frac{n}{N}\right) = \ln(\eta) - \alpha L_0 A_m(\theta). \quad (29.6)$$

By performing linear regression on the measured values x_i and y_i, the coefficients $\ln\left(\frac{n}{N}\right)$ and $A_m(\theta)$ can be fitted. Based on least squares analysis, the coefficients and their errors can be solved.

29.2.2 *Principle of Star Selection*

Before the measurement, it's necessary to choose suitable standard stars from the star library. The selection of standard stars should adhere to the following criteria:

(1) Star brightness falls within the observation capabilities of the telescope. The stars should not be too bright, as TianQin Station uses Superconducting Nanowire Single Photon Detector [7, 8] (SNSPD), and excessive photons can introduce errors in the number of photons detected by the detector compared to the actual photons reaching the detector surface.
(2) Spectral data provided by the star library should cover the system's working wavelength range. The optical system at TianQin Station operates at a wavelength of 1064nm, with all optical surfaces coated with coatings specially designed for 1064nm. Therefore, the selection of standard stars should encompass the 1064nm wavelength.
(3) During the measurement experiment period, the star's zenith angle should smaller than 90 degrees.

29.3 Experiment Results

TianQin Station [9–11] is equipped with a 1.2-m telescope, with the primary mirror (M1) having a diameter of 1.2 m and the secondary mirror (M2) having a diameter of 0.3 m. The optical structure of the laser ranging system is shown in Fig. 29.2. The echo receiving system consists of the telescope (M1-M3), the Coudé optical path (M4-M7), mirrors (BS & spin mirror), and the detector.

On the selection of the standard star library, the IRTF Spectral Library contains stars (including F, G, K & M stars) that meet the measurement requirements of TianQin Station. The IRTF Spectral Library is a collection of spectra primarily covering the 0.8 ~ 5.0μm range, mainly stellar spectra observed at a resolving power

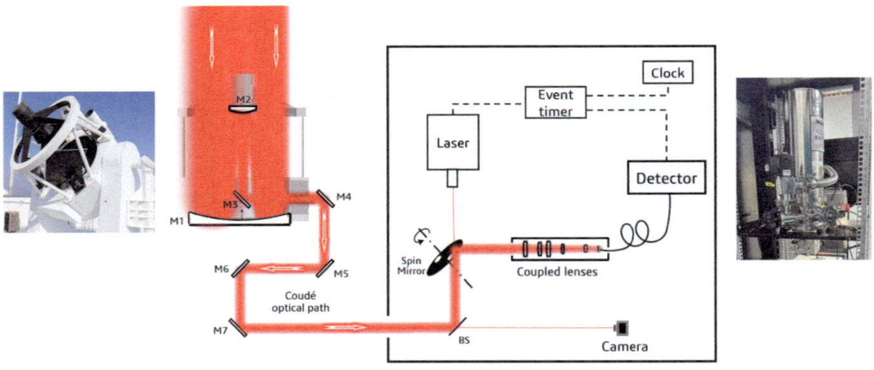

Fig. 29.2 Diagram TianQin Station system, the red beam indicates the echo receiving path

of R≡λ/Δλ ~ 2000 using the SpeX medium-resolution spectrograph at the Infrared Telescope Facility (IRTF) on Mauna Kea [12].

To validate this method, we conducted experimental measurements on two occasions: once (July 14th) before the mirror was cleaned, and the second to the third (Oct.24th & Nov.2nd) after cleaned the mirror (M1-M3, M5, M7, BS & spin mirror). During these measurements, the efficiencies of TianQin Station's echo receiving system was measured and compared.

On 2023/7/14 and 2023/11/2, experiments were conducted using TianQin Station's laser ranging echo receiving system. From IRTF star library, we have chosen a total of 18 stars meeting the star selection criteria mentioned in Sect. 0. Table 29.1 (column 1, 2, 3) provides the star name, flux, and their corresponding idealized photon flux (N) after conversion through Eq. (29.1). Based on the positions of the standard stars listed in the table below, the telescope was aligned with the standard star during the experiment. This alignment was confirmed by monitoring with a camera to ensure that the standard star was at the center of the field of view. The detector was then activated to measure the received photons from the star. Simultaneously, the current zenith angle position data of the star was recorded (Table 29.29. 29.2).

Table 29.1 Selected stars, their flux and the experiment results (2023/7/14)

Star name	Flux (Wm^{-2} μm^{-1})	N (Photons/s)	Zenith angle*	Airmass	n (Photons/s)
HD 126,660	2.49E-10	3.02E + 06	68.5°	2.70544	56,152
HD 135,722	8.97E-10	1.09E + 07	56.25°	1.79429	219,006
HD 137,759	1.16E-09	1.41E + 07	55.45°	1.75803	279,444
HD 142,091	2.57E-10	3.11E + 06	54.17°	1.70364	62,360
HD 161,664	1.57E-10	1.90E + 06	57.93°	1.87672	42,591
HD 176,051	1.05E-10	1.27E + 06	23.8°	1.09261	32,848
HD 181,475	2.12E-10	2.57E + 06	28.6°	1.13845	66,604
HD 187,238	1.41E-10	1.71E + 06	7.91°	1.00958	41,611
HD 192,713	1.62E-10	1.96E + 06	1.32°	1.00026	47,452
HD 194,193	2.10E-10	2.54E + 06	19.13°	1.05825	63,320
HD 204,724	8.66E-10	1.05E + 07	26.5°	1.11697	237,988
HD 212,466	2.07E-10	2.50E + 06	41.77°	1.3392	58,263
HD 213,306	2.96E-10	3.59E + 06	46.42°	1.44826	86,672
HD 213,893	1.00E-10	1.21E + 06	35.38°	1.22556	30,815
HD 215,648	2.13E-10	2.58E + 06	40.76°	1.31875	64,347
HD 216,946	5.97E-10	7.22E + 06	47.3°	1.47205	157,086
HD 219,134	1.05E-10	1.27E + 06	45.06°	1.4136	32,677
HD 219,734	8.88E-10	1.07E + 07	51.61°	1.6066	248,383

* Zenith angle: refers to the zenith angle of the star at the time of measurement

Table 29.2 Selected stars, their flux and the experiment results (2023/11/2)

Star name	Flux (Wm^{-2} μm^{-1})	N (Photons/s)	Zenith angle*	Airmass	n (Photons/s)
HD 10,465	3.15E-10	3.81E + 06	43.28°	1.3718	164,258
HD 137,759	1.16E-09	1.41E + 07	79.17°	5.14847	333,198
HD 142,143	1.38E-09	1.67E + 07	79.92°	5.50036	429,558
HD 164,136	2.86E-10	3.46E + 06	65.59°	2.40407	111,658
HD 164,349	4.01E-10	4.86E + 06	67.33°	2.5749	153,978
HD 181,475	2.12E-10	2.57E + 06	59.72°	1.97521	88,918
HD 182,835	2.67E-10	3.24E + 06	55.36°	1.75407	109,858
HD 192,713	1.62E-10	1.96E + 06	39.19°	1.28894	86,958
HD 204,585	1.30E-09	1.57E + 07	14.54°	1.03298	717,718
HD 204,724	8.66E-10	1.05E + 07	14.32°	1.03196	443,838
HD 212,466	2.07E-10	2.50E + 06	34.5°	1.21253	104,898
HD 213,306	2.96E-10	3.59E + 06	36.42°	1.2417	149,898
HD 214,665	1.11E-09	1.35E + 07	34.6°	1.21399	577,958
HD 215,648	2.13E-10	2.58E + 06	77.71°	4.57738	66,638
HD 216,946	5.97E-10	7.22E + 06	27.94°	1.13145	322,718
HD 219,734	8.88E-10	1.07E + 07	28.3°	1.13524	441,098
HD 220,657	1.66E-10	2.01E + 06	5.11°	1.00398	87,178
HD 23,475	2.32E-09	2.81E + 07	66.22°	2.463	921,458
HD 3346	4.61E-10	5.58E + 06	32.03°	1.17886	237,478

* Zenith angle: refers to the zenith angle of the star at the time of measurement

The plot of $\ln(n/N)$ against the airmass was plotted, as shown in Fig. 29.3.

The measurement result of the echo receiving system efficiency is: $\eta = 0.0304$ and $\eta = 0.0491$ (on 2023/7/14 and 11/2 respectively);

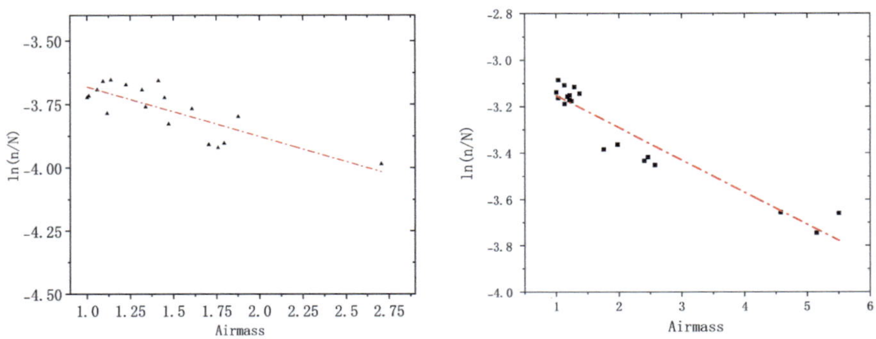

Fig. 29.3 The correlation between ln(n/N) and airmass along with its linear regression fits

We conducted calculations on the system efficiency using the parameters in manufacturing [10]: the optical system transmittance is 20%, and the detector efficiency is 30% (@2MHz). Consequently, the theoretical efficiency is 6%, showing a discrepancy of 1.2% from the measured value. Considering the estimation error and the efficiency degradation, we affirm the accuracy of our measurements.

29.4 Conclusion

Laser ranging system are indispensable for a wide range of applications, from satellite orbit determination to space debris tracking. And the performance of the echo receiving optical system, a vital component of laser ranging systems, directly affects the system's overall capability.

This paper has introduced a practical method for measure the efficiency of the echo receiving system. The method eliminates the need for additional devices and offers a cost-effective solution for measurement. It can be applied to ensure the capability monitoring of laser ranging systems. By measuring the photon flux at various zenith angles and compensating for atmospheric effects, we successfully experimentally measured the efficiency of the echo receiving system at TianQin Station, and the system's overall efficiency was 4.85% now. The results demonstrated the method's effectiveness and reliability.

References

1. P. Bing, Z. Fumin, Q. Xinghua et al., Design and analysis of echo receiving system in non-cooperative target ranging. Chin. J. Sci. Instrum. **1**, 198–205 (2017)
2. P.R. Spyak, W.L. Wolfe, Scatter from particulate-contaminated mirrors. part 4: properties of scatter from dust for visible to far-infrared wavelengths. Opt. Eng. **31**, 1775–1784 (1992). https://doi.org/10.1117/12.58711
3. S. Addanki, I.S. Amiri, P. Yupapin, Review of optical fibers-introduction and applications in fiber lasers. Results in Physics **10**, 743–750 (2018). https://doi.org/10.1016/j.rinp.2018.07.028
4. L. Xue, Z. Li, L. Zhang et al., Satellite laser ranging using superconducting nanowire single-photon detectors at 1064nm wavelength. Opt. Lett. **41**, 3848–3851 (2016). https://doi.org/10.1364/OL.41.003848
5. J.J. Degnan, Millimeter accuracy satellite laser ranging: a Review. AGU Geodynamics. Series, 133–162 (1993). https://doi.org/10.1029/GD025p0133
6. A.T. Young, Air mass and refraction. Appl. Opt. **33**, 1108–1110 (1994). https://doi.org/10.1364/AO.33.001108
7. Q. Zhao, A. McCaughan, F. Bellei et al., Superconducting-nanowire single-photon-detector linear array. Appl. Phys. Lett. 142602 (2013). https://doi.org/10.1063/1.4823542
8. J. Chang, I. Esmaeil Zadeh, Superconducting single-photon detectors get hot. Nat. Nanotechnol. **18**, 322–323 (2023). https://doi.org/10.1038/s41565-023-01334-1
9. J. Luo, L. Chen, H. Duan et al., TianQin: a space-borne gravitational wave detector. Class. Quantum Gravity **33**, 35010 (2016). https://doi.org/10.1088/0264-9381/33/3/035010
10. C. Zhang, T. Gao, Y. Cao et al., The facilities and performance of TianQin laser ranging station. Classical. Quantum Gravity **39** (2022). https://doi.org/10.1088/1361-6382/ac6d3e

11. T. Gao, L. Zhou, C. Zhang et al., Lunar laser ranging based on a 100Hz repetition frequency. Appl. Opt. **60**, 11058 (2021). https://doi.org/10.1364/AO.442263
12. J.T. Rayner, M.C. Cushing, W.D. Vacca, The infrared telescope facility (IRTF) spectral library: cool stars. Astrophys. J. Suppl. Ser. **185**, 289–432 (2009). https://doi.org/10.1088/0067-0049/185/2/289

MIX
Papier aus verantwortungsvollen Quellen
Paper from responsible sources
FSC® C105338

If you have any concerns about our products,
you can contact us on
ProductSafety@springernature.com

In case Publisher is established outside the EU,
the EU authorized representative is:
Springer Nature Customer Service Center GmbH
Europaplatz 3, 69115 Heidelberg, Germany

Printed by Libri Plureos GmbH
in Hamburg, Germany